Anonymous

The North Atlantic Telegraph Via the Faeroe Isles, Iceland, and Greenland

Anonymous

The North Atlantic Telegraph Via the Faeroe Isles, Iceland, and Greenland

ISBN/EAN: 9783337318345

Printed in Europe, USA, Canada, Australia, Japan

Cover: Foto ©berggeist007 / pixelio.de

More available books at **www.hansebooks.com**

THE

NORTH ATLANTIC TELEGRAPH

VIA

THE FÆRÖE ISLES, ICELAND, AND GREENLAND.

PROCEEDINGS

OF THE

ROYAL GEOGRAPHICAL SOCIETY OF GREAT BRITAIN,

JANUARY 28TH AND FEBRUARY 11TH, 1861.

REPORTS OF THE SURVEYING EXPEDITIONS,

ETC., ETC.

THE LORD ASHBURTON, PRESIDENT.

LONDON:
EDWARD STANFORD, 6, CHARING CROSS.
1861.

LONDON:
BRADBURY AND EVANS, PRINTERS, WHITEFRIARS.

CONTENTS.

		PAGE
I.	Introduction	7
II.	Meeting of the Royal Geographical Society, Jan. 28th, 1861 .	13
	Remarks by the Lord Ashburton.	14
III.	A paper on the Surveys of the Deep Seas, showing the practicability of the proposed North Atlantic Telegraph route; by Sir Leopold M'Clintock, late Commanding Her Majesty's ship "Bulldog."	16
IV.	A paper on the Surveys of the Landing-places for the proposed Telegraph, being a synopsis of the reports of Mr. Allen Young, Commander of the "Fox" Expedition; with other remarks by Sir Charles Bright . . .	29
V.	A paper on the Land Sections of the proposed Telegraph over the Færöe Isles and Iceland; by Dr. John Rae, Commander of the Land Expeditions.	44
VI.	A paper on the Fiords of South Greenland; by Mr. J. W. Tayler, late resident of Greenland for seven years. . .	61
VII.	Remarks on the Electric Circuits of the proposed North Atlantic Telegraph; by Col. Tal. P. Shaffner. . .	67
VIII.	Meeting of the Royal Geographical Society, Feb. 11th, 1861 .	75
	Remarks by The Lord Ashburton . . .	75
	„ Sir Edward Belcher, R.N.	76
	„ Pliny Miles, Esq.	78
	„ Captain Sherard Osborn, R.N.	81
	„ John Ball, Esq.	83
	„ Sir Roderick Murchison	85
	„ Dr. John Rae	86
IX.	Opinions of Residents of Greenland on the practicability of laying Telegraph Cables upon the coast of Julianshaab District	89
X.	Opinions of the Danish Commissioners on the practicability of the proposed Telegraph Route	94
XI.	Royal Danish Concession	95

MAPS, DIAGRAMS, AND ILLUSTRATIONS

EXHIBITED AT THE ROYAL GEOGRAPHICAL SOCIETY, PERTAINING

TO THE NORTH ATLANTIC TELEGRAPH.

1. A LARGE SPHERICAL MAP, embracing the seas and lands to be traversed by the proposed North Atlantic Telegraph; also of the Continents of Europe and North America.

2. A MAP, showing the rivers, lakes, geysers, mountains, glaciers, plains, farms, and towns of Iceland; also the track travelled over by Dr. John Rae, Commander of the Land Expedition of the North Atlantic Telegraph, and of the proposed route of the line across Iceland.

3. A LARGE MAP OF THE FÆRÖE ISLES, and another of the Stromoe Isle, showing the mountains, valleys, fiords, bays, and towns; and the track travelled over by the Land Expedition, and the route proposed for the Telegraph.

4. A MAP OF THE ATLANTIC OCEAN, with all the deep-sea soundings taken by the British and American Governments north of the Tropics, showing the superior favourable depths of the seas on the North Atlantic Telegraph route.

5. A MAP OF GREENLAND, showing the mountains, islands, unexplored regions, glaciers, and fiords, south of the Arctic Circle.

6. A Large Map of Julianshaab District, South Greenland, showing the mountain ranges, groups of isles, the many long, deep, and wide fiords in which the Telegraph Cables can be safely laid, and beyond the reach of icebergs.

7. Charts of the various bays and fiords examined by Mr. Allen Young, Commander of the "Fox" Expedition, and reported by him as suitable and safe places for the landing of the cables on the coasts of the Færöe Isles, Iceland, and Greenland.

8. Maps, with the deep-sea soundings, taken by Sir Leopold M'Clintock, on the proposed North Atlantic Telegraph route, viâ the Færöe Isles, Iceland, and Greenland; and with the soundings taken by Commander Daymon on the route of the late Atlantic Telegraph from Ireland to Newfoundland.

9. A Diagram, showing the mode of telegraphic manipulation over air lines, with the earth forming half of the circuit, connecting the respective voltaic batteries.

10. A Large Diagram, illustrating the mode of coupling together two or more electric circuits by mechanical contrivances, so as to overcome great distances by a combination of separate and independent circuits, charged respectively by batteries; precisely as can be practically applied and operated on the proposed North Atlantic Telegraph.

11. Model Stations, with telegraphic apparatuses, representing Europe, Færöe Isles, Iceland, Greenland and America; by which dispatches were sent by the motion of the finger from Europe to America, and simultaneously to the Færöe Isles, Iceland, and Greenland, precisely as can be done on the proposed North Atlantic Telegraph; so that dispatches need not be re-written at the respective places, but sent instantaneously from Europe to America, in the same manner as is daily done from London to St. Petersburgh, to Vienna, to Constantinople, and other places far distant.

12. A Chart of the surveys of Hamilton's Inlet and the sea adjacent thereto, by Sir Leopold M'Clintock's expedition, showing that

the water is sufficiently deep to permit a heavy shore cable to lie in safety.

13. ONE HUNDRED AND SIXTY STEREOSCOPIC VIEWS, illustrating the people, houses, churches, farms, towns, bays, mountains, valleys, water-falls, geysers, and various subjects of interest on the route of the proposed telegraph at the Færöe Isles, Iceland, and Greenland; taken by Mr. John E. Woods, of the "Fox" Expedition.

14. SPECIMENS of deep-sea soundings, taken on the route of the proposed telegraph, exhibited through microscopes. [These soundings have been examined by the philosopher Ehrenberg, and they prove his theory of life at the bottom of the ocean; and, also that the bottoms of the respective seas are highly favourable for the proposed telegraph.]

THE

NORTH ATLANTIC TELEGRAPH.

The idea of laying down a telegraph line between Europe and America, by what is now familiarly known as the North Atlantic *route*, was advanced many years ago both in England and America. In the year 1854 Colonel Shaffner of the United States took steps towards the actual realisation of the project. Owing, however, to the Russian war distracting public attention at the time, and subsequently to the engrossing claims of the attempt to carry a deep-sea cable direct from Ireland to Newfoundland, it was not until within the last twelvemonths that the British nation became acquainted with the proposed north-about route, as offering the most practicable mode of effecting Transatlantic telegraphic communication.

The success which attended the laying of the cable between Dover and Calais in 1851, gave an impetus to submarine telegraphy. Cables were laid from time to time over various rivers and arms of the seas; and as no instance of failure occurred, the possibility of crossing the Atlantic itself came to be regarded as a feasible project. There had been no manifestations in electrical phenomena to cause any doubt as to the successful transmission of the galvanic current through a cable of the length required to connect the two continents, a distance of 2100 miles, until the latter part of the year 1853, when it was made known by that distinguished philosopher, Professor Faraday, that the electric force is arrested in its transit through very long cables. Subsequent experiments demonstrated that this retardation is cumulative, increasing in a certain ratio in proportion to the distance traversed; that is to say, while it takes one-third of a second to transmit a current through a cable 500 miles in length, it will take about a second to transmit it 1000 miles. The discovery of this fact was considered by many electricians both in England and America to be a fatal obstacle to the commercial success of working a cable from Europe to America direct.

Among the most prominent who held this opinion was Colonel Shaffner of America. And early in the year 1855 we find that he published his opinion in the following language, which up to the present time he continues to entertain, viz.: " I will not say that a galvanic or magnetic electrical current can never be sent from Newfoundland to Ireland; but, I do say that, with the present discoveries of science, I do not believe it practicable for telegraph service." That gentleman had devoted himself to the science and art of electric telegraphy for many years, and had been the pioneer of the telegraph in the Western States, where under his guidance the wires were extended to the verge of civilisation in the West as early as 1851. By 1850 his name stood high as one of the first telegraphic engineers of the United States. A short time after Professor Faraday's lecture, he started the project of a telegraphic communication between the two continents by a series of short stages, in preference to a continuous line carried right across the ocean. His original idea was to connect Europe generally with America, and he accordingly proposed to carry his line from Labrador to Greenland, from Greenland to Iceland and the Færöe Islands, and from the Færöe Islands to Norway, and thence to Copenhagen. He visited Europe for the purpose of securing the necessary concessions, and in August 1854 obtained one from the King of Denmark, and others from the Governments of Sweden and Norway in the following year. Thus, at the time when the question of a connection by telegraph with America had been but scarcely entertained in this country, Colonel Shaffner had already laid down a route, availing himself of those natural stepping-stones which Providence had placed across the ocean in the North, and thus evading many of the difficulties and risks which it was predicted would be encountered in the direct sea route.

Subsequently, when the difficulties connected with that route came to be demonstrated by actual trial, Colonel Shaffner, with a view to meet the urgent wants of the British mercantile community, resolved upon diverting the European end of the cable from Norway to Scotland, which was cheerfully accorded by the Government of Denmark. In 1859 he took another important step in the execution of his project, which was to explore personally the intended route by Greenland and Iceland. He conceived he had good grounds for believing that the ice of Labrador and Greenland would not imperil the safety of a cable laid on the bed of the sea, provided the shore-end could be properly secured, and that there was nothing in the electrical conditions of the Northern regions to interfere with the working of the telegraph. He, accordingly, chartered a small sailing vessel, and embarking with his family on board, put forth from Boston on the 29th of August, 1859, for the purpose of making this preliminary survey. He landed at Glasgow in November of that year, and presented to the public the results of his voyage, at a meeting over which presided the Dean of Guild, Mr. Buchanan. The observations made and facts collected on that voyage gave the public reason to hope that telegraphic communication could be established by way of the proposed route, and so satisfied were his friends in London with the advantages held out, that funds necessary to secure the undertaking were promptly found.

During the voyage above referred to, Colonel Shaffner sounded the deep seas to be traversed by the telegraph between Labrador and Greenland, and between Greenland and Iceland. These and other observations have been most singularly corroborated by the more recent Government survey by Sir Leopold M'Clintock, and by that of Mr. Allen Young, commander of the " Fox " Expedition.

In the course of the spring of 1860, Colonel Shaffner read a paper on the proposed North Atlantic Telegraph to the members of the Royal Geographical Society, and every assistance was rendered, under the able direction of the President, Earl de Grey, Sir Roderick Murchison, and the Secretary, Dr. Norton Shaw. On the 15th of May, Lord Palmerston granted an audience to an influential deputation, headed by the Right Hon. Milner Gibson, and four other members of the House of Commons, to solicit the assistance of Government in sending out ships and officers to make the necessary official survey for ascertaining the practicability of the proposed route. The Premier appeared fully to appreciate the advantages of the north-about scheme, and in a very short time the Admiralty were directed to send out an expedition for the purpose of making the required survey.

The Admiralty selected for this duty Captain Sir Francis Leopold M'Clintock, an officer of great experience in the navigation of the Arctic Seas, and Her Majesty's steamer " Bulldog " was placed under his command. This distinguished officer was directed to take the deep-sea soundings, and he sailed from Portsmouth on his mission in June, 1860. In the meantime, the promoters of the enterprise purchased the " Fox," the yacht formerly employed in the successful search for the remains of the Franklin expedition, and fitted her out for the purpose of making the surveys of the landing places of the respective cables. The "Fox" was placed under the command of Captain Allen Young, of the mercantile marine, an officer well known for his distinguished labours under M'Clintock in the Franklin search. At the same time, Dr. Rae, an intrepid Arctic explorer, volunteered his services to join the "Fox," and take charge of the overland expeditions in the Færöe Isles, Iceland, and Greenland. Whilst Colonel Shaffner, as Concessioner and Telegrapher, and two delegates on the part of the Danish Government, Lieutenant Von Zeilau and Arnljot Olafsson, a member of the Icelandic Diet, accompanied the "Fox" Expedition, to take part in the necessary surveys.

Before the departure of the "Fox," which sailed on July 18th, 1860, Her Majesty the Queen, the Prince Consort, and other members of the Royal Family, honoured the enterprise by a visit to that vessel, while lying off Osborne, and took a lively interest in the details of the expedition.

The "Bulldog" returned to England in November, and the "Fox" arrived towards the latter part of the same month. Both expeditions, notwithstanding an unusually tempestuous season, were successfully carried out, and on the evening of Monday, the 28th January, in the present year, the results of their labours, as well as those of Dr. Rae and Colonel Shaffner, with a report upon the Greenland Fiords, by Mr. J. W. Tayler, a resident on that coast for seven years, were presented to the members of the Royal Geographical Society.

The respective papers were received with repeated applause, and every possible manifestation was given, indicating the confidence of those present in the practicability of the enterprise. At the subsequent meeting, February 11th, a discussion upon the papers took place, on which occasion the promoters and others engaged in this great undertaking were enthusiastically congratulated upon their most triumphant success, in proving the practicability of the proposed telegraphic route, *viâ* the Færöe Isles, Iceland, and Greenland.

Before closing these remarks it is but justice to notice the assistance given to the enterprise by Mr. Joseph Rodney Croskey, the eminent American merchant, of London. This gentleman not only advanced the £11,000 caution money paid to the Danish Government, but he also made the requisite advances for the purchase of the "Fox," and fitting her out for the late Expedition.

THE ROYAL GEOGRAPHICAL SOCIETY.

MONDAY EVENING, JANUARY 28, 1861.

A CROWDED meeting of the members and their friends assembled at Burlington House, on Monday, January 28th, to receive the Reports of the late Expeditions for the survey of the North Atlantic Telegraph route.

The LORD ASHBURTON, President of the Society, in the Chair.

Among the many distinguished and influential personages present were Sir R. I. Murchison, His Excellency the Danish Minister M. de Billé, Count Apponyi, Sir Walter Trevelyan, Admiral Sir Thomas Herbert, General Portlock, R.E., Lord Alfred Churchill, M.P., Sir Thomas Fremantle, Sir H. Maxwell, Sir John Login, Sir Charles Bright, Captains Collinson and Ommanney, Sir F. L. M'Clintock, Sir F. Nicolson, Sherard Osborn, J. Stopford, G. A. Bedford, Byron Drury, A. Ryder, and Pike, R.N., Hon. A. Kinnaird, M.P., Mr. J. Wyld, M.P., Mr. Cornelius Grinnell, of New York, Dr. Hamel, of Russia, Colonels Loyd and Cartwright, Major Edwards, R.E., Captains Sydney Webb, C. Johnston, R. Burton, Claude Clerk, E. M. Jones, and D. J. Herd, Drs. Ogle, Lister, Bennett, Bigsby, and M'Cosh, Messrs. J. R. Croskey, J. E. Davis, W. J. Hamilton, J. Crawfurd, Brooking, C. White, R. Bentley, J. Murray, C. V. Walker, George Saward, Arrowsmith, E. O. Smith, Cook, Gould, Vaux, Finlay, Rev. C. P. Wilbraham, and Colonel Shaffner, Dr. Holland, Pliny Miles, and Geo. F. Train, of the United States.

Sir Charles Bright, General Eber, the Rev. C. J. Fynes Clinton, Consul C. P. Hodgson, of Japan, the Rev. F. H. Mainwaring Sladen, the Marquis of Sligo, the Rev. C. Hill Wallace, the Rev. S. E. Wharton, Lieutenant A. S. Windus, I.N., and James Aiken, Walter Brodie, James Campbell, John E. Davis, R.N., N. Vaughan Edwards Vaughan, J. Wilson Holme, John Learmouth, Alexander Macmillan, George Newman, G. H. Oliphant Ferguson, Julius Reuter, Frederick Simpson, Ronald Thomson of Tehran, John Walker, and W. Warder were elected Fellows.

THE NORTH ATLANTIC TELEGRAPH.

The Lord Ashburton.—The subjects for our consideration to-day are of so interesting a character, that I should be doing unwisely if I were to occupy your time beyond a few moments, in making such preliminary observations as are expected from your President.

The papers which are about to be read relate altogether to the physical and geographical facts upon which the proposition has been based for the extension of the Electric Telegraph line between this country and America, by the route of the Færöe Islands, Iceland, Greenland, and Labrador. It comes especially within the province of the Royal Geographical Society to receive and to record those facts as purely belonging to the science of Physical Geography. We are so fortunate as to have a great many labourers working for us in the field of geography; some who have been sent out by the Society, like Speke and Livingstone, are doing our work, and at the same time doing work for many other societies. We have labourers also not less diligent, and not less useful to us as well as to the world: they are those who are occupied in other pursuits, but who come to us to record the facts they have observed, and to enable us to treasure them up as the material for future research.

Among the labours of this class, not the least valuable are the researches which we are met to record to-day. They were undertaken not for the purpose of acquiring geographical knowledge simply, but in order to carry out the great and beneficent scheme of connecting the two continents of Europe and America by means of telegraphic communication. Yet, at the same time that we receive these facts, we must take care not to suffer ourselves to be led away into that which is altogether out of our province, an attempt to pronounce any judgment upon the value of that scheme. That is a question for others to decide, not for us.

Therefore, whilst we ourselves may look upon the facts that may be presented to us with philosophic calmness, there may, on the other hand, be those who are so deeply interested in the material success of electric telegraph connection with America, as to be drawn into an eager contest to set up the merits of one scheme in preference to those of another. Should there be any gentlemen present prepared to carry these feelings into the discussion, I feel assured they will remember this, that it is those who are beaten that generally complain, and that the man who is the winner at chess is not the man to throw the pieces at his adversary's head. I believe we shall have no symptoms of distress exhibited; that we shall not have anybody manifesting the conviction passing in his mind that he is worsted in the argument.

I will now call upon Sir Leopold M'Clintock to read the first paper that is set down on the list.

SURVEY OF THE DEEP SEAS,

AND OTHER OBSERVATIONS; BY

CAPTAIN SIR F. LEOPOLD M'CLINTOCK, R.N., LL.D., F.R.G.S.,

LATE COMMANDING HER MAJESTY'S SHIP "BULLDOG."

Sir Leopold M'Clintock then read the following paper:

In compliance with a request from the promoters of the North Atlantic Telegraph Route, her Majesty's Government dispatched the "Bulldog" under my command, on the 1st of July last, with orders to ascertain the depth of the ocean, and as far as possible the nature of the bottom, between the Færöe Islands and Iceland, Iceland and Greenland, and between Greenland and Hamilton Inlet on the Labrador coast. I was also directed, should my time permit, to make a slight examination of that inlet—being British territory;—but in no other instance did my duty extend to the examination of any of the coasts I was required to approach.

With the exception of Hamilton Inlet, none of the positions for the shore-ends of the proposed lengths of cable were suggested when I sailed from England. The duty of selecting them was subsequently entrusted, by the promoters of this Telegraph Route, to Captain Allen Young, in the "Fox"; consequently, my lines of soundings have not in every instance been carried in from the deep sea, so as to unite exactly with the coast explorations of Captain Young.

Although my visit to the Færöe Islands was not for the purpose of making any examination of their shores, yet I could not fail to observe that a submarine cable, in connection with the main island, and a land-wire across it, could be maintained with perfect ease.

In my official report to the Secretary of the Admiralty, written previously to my return to England, and which I shall have frequent occasion to quote, I have remarked that on landing at Thorshaven, the chief town of the islands, I observed that the little bays near it afforded ample shelter and security for any cable landed within them.

The best harbour in the group is Westmanshaven, but it is situated in a channel through which the tide runs fully six miles an hour, and for this reason it would not be advisable to bring the cable there.

I was informed that the channel between the islands of Stromöe and Osterðe is almost obstructed in the middle, being contracted to fifty or eighty yards; hence there can be but a very slight flow of tide through it, and upon this account I would seek a landing-place for the Iceland cable near to the north-west outlet of this channel, at Haldervig or Eide.

Leaving the Færöe Isles on the 6th of July, we sounded across towards Ingolfsholde upon the S.E. shore of Iceland, a distance of 280 miles, and found the depth to be generally less than 300 fathoms, the greatest depth being 680 fathoms. The specimens of the bottom consisted, chiefly, of fine sand, or mud and broken shells, and in two instances, of minute volcanic debris; the temperature of the sea at 100 fathoms below the surface scarcely varied from 46°.

The depth of water upon this section of the telegraph route is so moderate, that it would be an easy matter to lay down a cable between Færöe and Iceland.

Since my return I find that Bern Fiord, upon the east coast

of Iceland, has been examined with a view to its selection as the landing-place for a cable; it is about eighty miles to the north-east of Ingolfsholde, and has the advantage of being somewhat nearer to Færöe.

On the 11th of July I arrived at Reikiavik, the chief town of Iceland; an expected supply of coals had not arrived, therefore I remained only three days, but returned again in October, when my stay extended from the 19th to the 28th. During these visits I obtained some interesting information about its physical aspect, its climatic condition, and the movements of the ice in the adjacent seas. I was informed that a telegraphic wire could not be carried along the south shore eastward of Portland, on account of the many wide rivers which have their sources amidst the mountains and glaciers of the interior. These rivers are much swollen in spring, when they carry down vast quantities of ice, and sometimes change their beds; but to the north of the central mountains no such difficulty would be experienced.

The east and west coasts are very seldom visited by drift-ice, not oftener than seven or eight times in each century, whilst it is only upon two or three of these occasions that the drift of Arctic ice is sufficiently extensive to reach the south coast. True icebergs are *never* seen; the masses sometimes mistaken for them are small enough to float in comparatively shallow water, so that a cable would remain undisturbed at the bottom, its shore end being carried into a fiord. Faxe Bay, on the south-west coast, enjoys a remarkable exemption from drift-ice; the last mention of its appearance within it is as long ago as 1683; neither does it freeze over—merchant vessels trade there throughout the winter. A cable could therefore be landed in this bay with perfect ease and security, and probably to the westward of Reikiavik.

The entire population of Iceland scarcely exceeds 60,000

souls. Education is perhaps more generally diffused than in any other country, and the topographical maps recently published by the Danish government delineate its features most fully, and with the greatest possible accuracy, and would greatly facilitate the survey of a land-line.

Although Iceland is considerably larger than Ireland, and is of volcanic origin throughout, yet for long ages the disturbance occasioned by its subterranean fires has been limited almost exclusively to its south-western quarter, where Hecla is occasionally, and Katla has been very recently, in an active state, and where Geysers and boiling springs are numerous; nor is the adjacent sea free from like convulsions. In 1783, a submarine volcano burst forth in a probable depth of 200 fathoms, about thirty miles off the S.W. extreme of the island; by it a new islet was formed; it soon after subsided, but still exists under water as a dangerous sunken rock. This volcano was again active in 1830;* its action appears to have been very limited, and within four leagues of it stands the time-honoured "Grenadier's Cap," a basaltic column, eighty feet above the sea; within 500 or 600 yards of this most remarkable rock the "Bulldog" sounded in seventy fathoms.

Fortunately the telegraph route is not required to pass, by sea or land, through any part of this disturbed or suspected area.

Five days of very calm weather enabled us to complete the line of soundings between Faxe Bay and the south-east coast of Greenland.

The depths generally were very regular, the greatest being 1572 fathoms, and situated in mid-channel; but when within forty miles of Greenland the depth decreased from

* Some interesting notices of this and other submarine volcanoes are published in the "Nautical Magazine," for July, 1860.

806 fathoms to 228 fathoms, in the short distance of 3¼ geographical miles.

The nature of the bottom was chiefly ooze, that is fine mud partly consisting of minute organic remains, but near to Iceland, volcanic mud and sand was more frequently brought up.

The temperature of the sea at 100 fathoms below the surface gradually diminished from 46° near Iceland, to 39° off the Greenland coast.

Circumstances which it is unnecessary to allude to here prevented me from commencing, before 18th August, the line of soundings between the S.W. coast of Greenland, and Hamilton Inlet on the Labrador coast,—a distance of 550 miles.

The Greenland shore was still blockaded by such a vast accumulation of drift-ice that we could not approach within 45 miles of it, at which distance the depth was ascertained to be 1175 fathoms. This line of soundings to Hamilton Inlet shows that the greatest depth,—which is in mid-channel,—is 2032 fathoms; and that the decrease is very gradual until within about 80 miles of Labrador, where there is a change from about 900 fathoms to 150 fathoms in 7 or 8 miles.

The ocean bed consisted of ooze, but with fewer microscopic organisms than previously met with, whilst the average temperature of the sea at 100 fathoms below the surface was 40°.

Seven days were all I could devote to the examination of Hamilton Inlet. Its length was found to be 120 miles, whilst its width varies from about 15 miles at its mouth, to scarcely half a mile at " the Narrows," which are about half way up to its head, and above which it expands into an inland sea of about 20 miles in width. All this great inlet was rapidly explored, its main channel from " the Narrows " to seaward was sounded, and the whole laid down by Mr. Reed, master and assistant-surveyor, with sufficient accuracy for ordinary purposes; but these soundings are not nearly sufficient to meet the requirements of a cable-route, nor even to decide whether a cable should be landed there.

We found the depths to be very irregular, and seldom sufficient to secure a submerged cable from disturbance by icebergs, A perfect survey is absolutely necessary, and may show that the shallow water and reefs of rocks which to our imperfect knowledge appeared intricate and unfavourable, may not only be avoided, but may afford a sure protection against the intrusion of icebergs within the mouth of the inlet.

There are some small rocky islets off the mouth of this inlet, and of these the Hern Islets lie nearly in the middle and contract the widest channel of entrance to about 5 miles; the greatest depth obtained in this channel was 49 fathoms. Had the depth of water amounted to 70 fathoms in as far as this position, I would not hesitate in pronouncing favourably of Hamilton Inlet as a terminus to the cable from Greenland.

The greater part of the local information which I obtained here was kindly furnished by Captain Norman, a Newfoundland merchant who has traded here each successive summer for 24 years; during the summer he resides at Indian Harbour, at the north entrance of the inlet, where there is a secure anchorage for vessels of moderate size. Captain Norman states that icebergs very rarely enter the mouth of Hamilton Inlet, and never pass within the Hern Islets; and for these reasons: 1st, that the current which has borne them from the north is here deflected off-shore by the Esquimaux Islands, and carries them past the mouth of the inlet; and 2ndly, that the flow of water caused by the discharge of several large rivers into the inlet, still further aids in carrying the drift-ice and icebergs out to seaward.

During winter and spring this drifting ice prevents all access to Labrador, but by June, Hamilton Inlet is usually quite free from it.

From Captain Norman I also learnt that the deepest water along the coast is off Cape Harrison, and that a large river runs into Byron Bay adjoining it; moreover, Sloop Harbour (which is close to the river) is said to be an excellent one. Unfortunately my time was too limited to admit of any examination of this promising locality. It is very desirable to obtain more information respecting the ice and icebergs upon this coast. It could be furnished by the Newfoundland traders and seal-fishers, and perhaps by persons in the employ of Messrs. Hunt, Henley & Co., of 8, Broad Street Buildings, E.C., a firm which has maintained an extensive establishment near to Hamilton Inlet for a very long period, fifty or sixty years I believe. In addition to these sources of information, there are intelligent Moravian missionaries, whose settlements on the Labrador coast have existed for more than one hundred years.

The shores of Hamilton Inlet appear bold, rocky, and almost devoid of vegetation when viewed from the sea; as we advance up it, the land becomes lower, the undulations more gentle, verdure and trees appear, and at its head the whole country is densely covered with spruce, white pine, and white birch, but the tallest trees do not exceed forty feet.

I was informed that the interior is similarly wooded, and has an exceedingly scanty population of Indians, allied to the Cree nation; they all profess Christianity, and are a strictly honest quiet race.

The residents along the shores of this great inlet are of European or mixed blood, and do not amount to 200 souls. During summer they catch codfish, herrings, and salmon; and in winter they are occupied in trapping fur-bearing animals.

At the Hudson Bay trading post upon North-west River, at the head of the inlet, I met Mr. Smith, the gentleman in charge, who kindly supplied me with the only information respecting the interior that I was able to obtain. He seemed to think there would be no difficulty in carrying a wire from here overland to Mingan, on the Gulf of St. Lawrence. The Indians frequently travel from one place to the other, the distance not exceeding 250 miles. Should the cable be taken to this inlet, I would suggest that it be landed upon the south shore, to seaward of "the Narrows," as the tides run through them with very great velocity. All other parts of the inlet freeze over to a depth of three feet, for the winters are very severe. The summers, though short, are no less remarkable for their warmth. At Northwest River barley and oats ripen, and potatoes and other vegetables grow tolerably well. Mosquitoes are such an intolerable plague, especially to new comers, that unless their faces are carefully veiled or smeared with camphorated oil, brimstone ointment, or dilute creosote, they cannot either repel or endure their bloodthirsty attacks.

Leaving Labrador on the 17th September, I returned to Greenland for the purpose of completing such soundings as the drift ice had previously compelled me to leave undone. Being, moreover, very desirous of meeting the "Fox," and of ascertaining from Captain Young where the cables were to be landed, so that I might continue the deep-sea soundings in to those positions, I visited the settlement of Julianshaab on the 29th September, but no information could there be obtained of the "Fox."

The season was very remarkable for the great quantity of drift ice which encumbered the shore, and had hitherto prevented vessels from approaching Julianshaab; in fact, so much ice had not been known for nearly 30 years.

This coast, I may remark, is usually quite free from ice by September.

Following up my inquiries, I learnt that the climate is not

nearly so severe as is generally supposed, the fiords are only partially frozen over in winter; a few cows, goats, and poultry are reared: and although the summers are cold, turnips, spinach, lettuce, and radishes grow in the open air.

I was informed that the large fiord of Tessermiut, which lies midway between Julianshaab and Cape Farewell, was the most likely place to afford security for a cable: that icebergs never came into it, and that there would be found ample depth of water from it out to sea; also that there is safe anchorage in a spacious bay near its mouth as well as high up in the fiord.

On 3rd October I put to sea, intending to sound into Tessermiut Fiord should the ice permit; but it was with difficulty we got out, for a S.E. wind had brought up much more ice from Cape Farewell, and prevented our approaching within 40 miles of Tessermiut or the adjoining coast; and the ship sustained considerable damage from unavoidable collisions with the ice before she got clear out to sea.

It is well known that a current from the North Atlantic Ocean bears along with it all this ice round Cape Farewell, and up the west coast of Greenland for several hundred miles. It carries the drift-ice for the most part along the outer islands, and it is only when there is a strong wind blowing in from the sea, that the ice comes in between the islands and enters the fiords; it is almost exclusively low or flat ice which thus drifts in, the larger masses and icebergs, which draw more water, nearly always keep themselves in the main stream along the outer islands.

It is evident, that were a cable brought in from the deep water existing outside and between these islands, and carried sufficiently far up a deep fiord, its security from icebergs would be insured; and that to protect the mere shore-end from the ordinary flat ice would be a matter of no great difficulty.

Since my return to England I have received a letter from the Resident Inspector of South Greenland, the well known Dr. Rink, whose writings on Greenland have added so largely to our knowledge of the physical condition of that great arctic continent. The opinion of such a man deserves serious attention, since it is scarcely possible to quote a higher authority upon the point in question. I therefore do so almost in his own words.

"I have thought much," he writes, " over the proposed route for the North Atlantic Telegraph ; at first I doubted the possibility of accomplishing it, but now I am of a contrary opinion. You can lay down the cable from Iceland round Cape Farewell into some fiord upon the south-west coast, where ice cannot ground upon it, or touch it except for a few fathoms out from the shore, and this last part may be easily protected. But to carry the wire across the interior of Greenland, as I have heard of, would be impracticable." This letter was written in Greenland, before either the "Bulldog" or "Fox" had arrived there, and experience has since shown the necessity for acting in accordance with the suggestion of Dr. Rink. The length of cable required to unite Iceland with West Greenland will be about 800 miles.

Finding that nothing more could be done upon the Greenland coast, I commenced a line of soundings towards Rockall, but a succession of tremendous storms and want of fuel prevented the completion of this service. One of the few casts obtained deserves particular mention ; the depth was first ascertained to be 1260 fathoms : then a sounding machine was lowered to obtain a specimen of the bottom, and about fifty fathoms of line more than the depth required was payed overboard, to insure its being down. On hauling it in, several small star-fishes were found adhering to that part of the line which had lain upon the bottom! The nearest land at the time was Iceland, and it was 250 miles distant. I simply mention this interesting fact, which I witnessed, leaving it to be enlarged upon by Dr. Wallich, the able naturalist of the Expedition, who is still employed by the Admiralty in the microscopic examination of our specimens of the sea-bottom. The result of his investigations (which will be published hereafter) may be of great importance to Marine Telegraphy, proving, as it will do, the existence of animal life at very great depths.

We are aware that the coating of a Mediterranean cable was attacked by minute creatures allied to the ordinary Teredo,[*] at the depth of sixty or seventy fathoms, and should it be found that similar boring animals exist in great depths, it will become imperative to protect the insulation of the wire against their ravages; but time does not admit of a digression from the object of this paper, which is simply to lay before you my experience and opinion with regard to the physical aspect of the proposed route; it may not, however, be out of place to mention, that the great pressure exerted at depths approaching to 2000 fathoms is sufficient to squeeze the tar freely out of rope: could we recover a cable from these depths, we would find the tar similarly expressed from its canvas wrappings. If the tar used were of a sufficiently viscid description to harden and remain coated upon the wrappings, it would probably afford quite a sufficient protection against these destructive creatures.

Once laid in deep water, the North Atlantic Cable will probably be more secure and more durable than any other; as it will lie at the bottom of a sea where the temperature is unusually low, and where animal life is proportionately rare.

---

[*] See a Paper by J. Gwyn Jefferys, F.R.S., on "The British Species of Teredo," published in "The Annals and Magazine of Natural History," for August, 1860.

If, during the coming summer a final selection and survey of a landing place in Greenland be made, all that will remain to complete the entire route will be a landing position in Labrador; and that a cable can be safely landed upon some part of this coast, if not in Hamilton Inlet, it is hardly possible to doubt.

Judging then from my own experience, and from the facts which the voyage of the " Bulldog " has brought to light,—many of which are supported by the most reliable local authorities,—I am of opinion that with regard to the practicability of laying a North Atlantic cable, there are no grounds for serious misgivings; on the contrary, nearly all the information which has so far been ascertained is of a kind favourable to the accomplishment of the undertaking.

That there is usually impenetrable ice upon the south-west coast of Greenland for eight months out of the twelve,—(*i.e.* from January until September,)—we are well aware; and hence originates the chief difficulty of the route. It is obvious that the Greenland cables cannot possibly be laid down whilst this ice remains upon the coast; but in ordinary seasons it does not clear away until Autumn is far advanced, and stormy weather becomes frequent. This difficulty, I apprehend, however, is not an insuperable or extraordinary one, since it is common to all similar operations at sea, requiring for their accomplishment a like period of four or five consecutive days.

I have assumed that the ice ceases to obstruct the south-west shore of Greenland about the middle of September; but we can no more predict its movements, than we can foretel the temperatures of the seasons, and the winds by which those movements are governed.

Exceptional seasons occur when it would be imprudent to attempt laying a Greenland cable: also rare seasons when it could be laid as early as July: again, there are seasons when the icedrifts are detached from each other so that vessels watching their opportunity may freely pass, into harbour, or out to sea, during the summer months.

In order to meet these ever-varying circumstances, it is the more necessary that the utmost caution be observed in all matters connected with the laying down of the Greenland lengths of the great cable; that the most suitable steamers be selected, and the highest engineering and nautical skill be employed.

And that this country possesses all the needful appliances, and the amount of professional talent requisite for the accomplishment of this great undertaking, is no more to be doubted, than that she possesses men of sagacity to appreciate its vast utility, and of commercial enterprise to bring about so desirable an issue within the next two or three years.

SURVEY OF THE LANDING-PLACES

FOR THE PROPOSED CABLES; REPORTED BY

ALLEN YOUNG, Esq., F.R.G.S.,
COMMANDER OF THE STEAM YACHT "FOX" EXPEDITION;

WITH REMARKS BY

SIR CHARLES BRIGHT, F.R.A.S., F.R.G.S.

The LORD ASHBURTON.—The next is Sir Charles Bright's paper upon Captain Allen Young's Reports, which will be read by Mr. Galton, viz.:

I have been requested by the promoters of the North Atlantic Telegraph to present to the Royal Geographical Society, a synopsis of the report which has been handed to me by Captain Allen Young, upon his recent voyage in the steam yacht "Fox," and his careful and elaborate survey of the proposed telegraphic route between Europe and America, by way of the Færöes, Iceland, and Greenland. The history of this important undertaking, up to the present date, will probably be familiar to the minds of most persons present, and I will therefore commence with the departure of the "Fox" from Southampton, on the 19th June, for the Færöes, where she arrived on the 3rd of August.

FÆRÖE ISLANDS.

This most interesting group of isles, the capital of which is Thorshaven, lies some 200 miles north of Scotland, and is under the authority of the Danish Crown. I will not occupy the time of the Society in discussing the political, physical, or other characteristics of these islands, but proceed at once to quote some interesting extracts from Captain Young's report. He says:—

"We were naturally anxious to reach the spot at which our work was to commence, and to ascertain the first foreign station at which the telegraph cable was to be landed. We were glad to make the Faröe Islands, distant 50 miles, on the evening of the 2nd August, the remarkable clearness of the atmosphere and the height of the land making our distance from it apparently far less than we were by our observations. When 46 miles E.S.E. from Naalsöe we obtained soundings in 102 fathoms, sand and shells. We here passed through many patches of discoloured water of a reddish hue, caused probably by minute animalcules on the surface, specimens of which, brought up by the towing net, being preserved. Specimens of water, both on the surface and at various depths, were frequently obtained during our voyage and preserved, the temperature and specific gravity being registered in the meteorological journal.

On the morning of the 3rd the land was obscured by clouds and mist, which, as the sun rose, gradually dispersed, and enabled us to obtain views of the land, and also to fix our positions by Born's chart to commence a line of soundings into the north point of Naalsöe; the depths were 36 to 26 fathoms, with a bottom of sand and shell."

THORSHAVEN.

"On rounding the north part of Naalsöe we took a fisherman on board as pilot, and at 10·50 anchored in Thorshaven, and immediately commenced an inquiry and examination of the locality, and testing the accuracy of all the charts and maps in our possession.

The results were as follow :—Thorshaven and bay is protected by Naalsöe, and is land-locked, excepting on two points to S.E., and on one point to N.E. A swell sets in to the inner harbour with S.E. gales, but this cannot be to any very great extent, from the fact that vessels lie at their moorings throughout the winter. The bay has good anchorage, varying in depth from 25 to 8 fathoms, bottom of sand, gravel, and shells, with a few patches of hard ground. Vessels usually moor in the two inner harbours or creeks, the northern being most frequented for the facility it offers for loading and discharging cargoes. Either of the inner harbours would do very well to land the telegraph wires, but from the many vessels frequenting the port it appeared desirable to select another place, for even were the cable to be buoyed, the risk from the ships' anchors would be considerable, on account of the want of space—but half a mile southward of Thorshaven is a small cove called Sandygerde, where the cable could be landed in safety and clear of ships' anchors. This cove is 1½ cable's length across and about the same depth, and shoals gradually to a sandy beach; it is intersected at the head by a water-course and mill, the land sloping gradually up an extensive valley to the interior. As many additional soundings were obtained across the fiord as our time would admit, proving that although the channel is uneven there is nothing to prevent bringing a cable in from sea. From the most reliable information from pilots and our own observations, the stream on the flood never exceeds 4 knots on the strongest spring tides, whilst on the ebb it is much weaker, and at times scarcely perceptible. It is high water at full and change at 4 o'clock, the flood runs to the southward. The gulf stream appears to sweep round these islands from left to right, or direct as the hands of a watch, and therefore in sailing from Thorshaven for the northward, by starting with the first of the flood and passing to the southward of Stromöe and through Hestöe, and Westmanshaven Fiords, you can carry a 9 hours favorable tide. The rise of tide at Thorshaven does not exceed 6 feet."

Westmanshaven.

"We left Thorshaven at 1 p.m. passing through Hestöe and Westmanshaven Fiords, and anchored in Westmanshaven in the evening. The scenery in these fiords is very magnificent, and as we steamed through with a strong head wind and weather tide, the surface of the water covered with sea-birds, the lofty hills on either hand rising to the height of 1500 to 2000 feet, with vast basaltic caverns and columns in the cliffs, formed a picture not easily forgotten. As I had heard that Sir Leopold McClintock had already examined this port, I did not deem it necessary to delay the ship for that purpose. The fiord appears clean and clear, with deep water close into either shore. I was informed that there is 70 fathoms water in the middle, a little north of Welbestad, and the stream in strongest spring tides runs six knots through the fiord. The rise of water is much influenced by the winds outside—it has reached ten feet at spring tides, and has been known as low as four, but the mean rise appears to be from six to eight feet. Westmanshaven is said to be the best harbour in the islands; it is completely landlocked, with a bar, probably formed of the débris washed down from the surrounding hills, and accumulated by the action of the streams and eddies in the fiord. I fear the current in this fiord would be disadvantageous."

Haldervig.

"We left Westmanshaven on the evening of August 5th, and, after weathering the northern extremity of Stromöe entered the sound between Stromöe and Osteröe, and anchored at Haldervig at 11·30 p.m. on the same night. An examination of the port and estuary of the sound was commenced. The results of these observations, which occupied two days, were, that little or no stream is found in the sound, that Haldervig has good anchorage, and is perfectly landlocked; the deepest water is 34 fathoms, bottom black mud and sand, but that a sandbar exists between Eide Point and Stromöe, over which there are $8\frac{1}{4}$ fathoms in the deepest part; and as in northerly gales the sea is said to break upon it, I consider that the cable would require a strong shore-end to ensure its safety in crossing this place. This bar lies rather within the entrance and narrowest neck of the sound." In the summary, Captain Young states: "At Haldervig we surveyed harbour and fiord, and found all satisfactory, and I think that place to be well adapted for the reception of the cable. We found but little current, and the cable can be taken in, in a tolerable depth of water, into a perfectly land-locked position."

ICELAND.

"On approaching the coast of Iceland we got occasional soundings towards Ostré Horn, under which we were obliged to anchor in a dense fog, after getting inside an extensive and dangerous reef of rocks, called by the Icelanders the Hartinger and Bortinger. These reefs lie two miles east (true) off this cape. They do not appear on the Danish surveys, but I afterwards found them as a single rock upon a French chart."

Berufiord.

"On the morning of August 12, the fog having lifted, we weighed under steam, and got into a position to carry a line of soundings into Berufiord, between the islands of Papey and

Kogar Point — these soundings average about 30 fathoms, principally sand and shells. We anchored off Djupivogr factory the same day, and it being Sunday, we ceased operations during the afternoon. The weather that day was the finest we had had since we left England, and the evening was truly summer-like. During the following five days, and when not prevented by the prevalent rain and fogs, we proceeded with the examination of the fiord, and finding it would not be advisable to carry a cable into the small harbour of Djupivogr, on account of many rocks in its vicinity and its being the anchorage of the small vessels frequenting the coast, we sought for a more suitable landing-place higher up the fiord, and succeeded in finding an excellent bay, called Gautavik, on the north shore, five miles from the entrance. A depth of near thirty fathoms can be carried in from sea to within a quarter of a mile of the shore, while the bay itself afforded good protection and anchorage for any large ships that might be employed in the undertaking.

High water at Djupivogr at three o'clock, full and change, rise six feet. The tide has been known to rise six and a-half feet before the coming of easterly gales; about the same time flood outside runs S.S.W. (true), ebb N.N.E., between Papey Island and the main. The strongest known stream has four knots, but the average in ordinary spring tides is not more than two and a-half knots.

In 1860 drift ice appeared off the coast and entered the fiord, and again (though in very small quantities) in 1859-60. This ice, called here Greenland ice, is the ordinary washed and decayed floe-ice, and comes from the N.W. *No icebergs have ever been seen on the coast.* The drift-ice appears with northerly and departs with southerly winds, and less of it comes into Berufiord than any other fiord on the east coast of Iceland; the residents accounting for this fact by Berufiord having a

S.W. direction, and is consequently protected by the more northerly and projecting capes which shunt the ice off, while the local tides keep it drifting up and down the coast. The fiord itself never freezes, but thin ice has been known to cover the harbour off the Factory for a day or two during the winter.

A tolerably complete survey of the fiord from the entrance to Gautavik was completed, but a further examination would be advisable outside, to ascertain the proper channel in which to lay the cable. The greatest difficulties experienced on the coast by seamen, are from the prevalent fogs during the summer months, and with easterly winds, *and this would render it advisable to start from this coast towards Faröes*, in laying the cable, because making a good land fall here would be attended with considerable uncertainty." Finally, as to the practicability of Berufiord, Captain Young says, " There will be no difficulties from the sea, ice, or otherwise, and the only obstacles will be from fogs and thick weather, but which may be overcome by selecting proper seasons, and taking precautions in landing or embarking the telegraph cable."

REIKIAVIK.

Captain Young sailed from Berufiord on the 17th day of August, and arrived at Reikiavik, the capital of Iceland, on the 21st day of August, and after making enquiries as to the coasts, he says, " I then determined to examine Hvalfiord, as from its situation it appeared to have the advantage over any place in Faxe Bay, and on the 27th I proceeded up that fiord, sounding it as far as ' Maria Havn,' a small harbour and salmon river on the south shore, 7 miles from the entrance of the fiord. The least depth of water in the channel of the fiord is 14 fathoms, with deeper water both outside and in, the general depth being 18 to 20 fathoms, soft mud. The cable could be taken into Maria Havn through soft mud, on a sandy beach in a land-locked position. Hvalfiord is protected from a heavy sea breaking into it by the shoals of ' Vestrhram ' and ' Sydiahraun ' in Faxe Bay, and on which there is less water than in the shoalest part of the channel of the fiord. The bays in the fiord are sometimes covered with thin ice, but the fiord itself never freezes, and with reference to drift-ice on this part of the coast, I cannot do better than quote the words of Sir Leopold M'Clintock. ' Faxe Bay never freezes over, and I can find no record of drift-ice within, since 1683. Merchant vessels come and go throughout the winter.' "

GREENLAND.

The "Fox" left Reikiavik, August 31st, and after a very rough passage, arrived at Frederickshaab, October 2nd. Captain Young remained there to make some necessary repairs, and finally arrived at Julianshaab on the 22nd October. He then reports : " Having made all inquiries about Igalikko or Julianshaab Fiord, I deemed it advisable at once to commence a survey of this beautiful arm of the sea, and acting upon the opinion of Colonel Shaffner, that were this fiord found practicable, the electric circuit from Reikiavik would not be too extended."

Julianshaab Fiord.

" We first sounded up to the head of the Fiord, which gave an opportunity for our landing a travelling party under command of Dr. Rae, to examine the inland ice and nature of the country. A party also went to the Old Nordisker Ruins, at Igalikko." Returning with the " Fox" to Julianshaab, October 27th, Captain Young then surveyed the estuary of the fiord, and from the soundings obtained, says, " I am of a decided opinion that *a* depth of not less than 150 to 160 fathoms can be carried from the middle of the fiord abreast the settlement, out to sea, with a general muddy bottom.

This depth of water will effectually preclude injury to the cable from the largest icebergs ever seen upon the coast. Although many bergs lay along the coast, we saw none aground in this valley of the fiord, nor according to information obtained from the residents, have they been seen grounded in that channel." Captain Young then proceeds to say—" This report and my previous letters will show that my decided opinion (so far as we have been upon that route) is favourable to the practicability of the undertaking, and that Julianshaab will, under all circumstances, be well adapted for the reception of the cable. With regard to the operation of laying the cable, I consider that no apprehension may be felt on that point, for from the sudden disappearance which we witnessed of the ice from the coast, and from the ice *usually* dispersing from the south-east shores of Greenland in the autumnal months, opportunities will always occur when a ship having the cable on board and lying in readiness in Julianshaab, may depend upon having a period of clear and open sea. *The cable once laid, no drift ice can in any way injure it, if the proper precautions are taken in securing the shore end.*"

Ice of the Greenland Seas.

" Since my arrival I have seen the admirable remarks of Mr. J. W. Tayler upon the southern coast of Greenland, the results of his experience during seven years residence there. His opinions must be most satisfactory to you, and I am sure that all who are interested in the work must be grateful to him for having so freely given them.

I perfectly coincide with his views with regard to the size of the icebergs frequenting the above coast and accompanying the Spitzbergen drift ice; and as this bears upon my own opinion that no iceberg will ground in the channel of Julianshaab Fiord, I think I may here explain my reasons for this statement. Having navigated the entire west coast of Greenland and into all the principal settlements, and having experienced a whole winter's drift in the ice, through Baffin's Sea and Davis' Strait, I have had occasion to remark and to gather all possible information upon the ice movements.

Around the coast of Greenland, westward of Cape Farewell, there are two distinct descriptions, or rather kinds, of drift ice ever approaching but never meeting together. The first is the ice formed during the winter on the vast area of Baffin's Sea and the different channels from the Polar Seas westward of Greenland. This ice, called by the Greenlanders *the west ice,* often blocks up throughout the year the upper part of Melville Bay, and drifts constantly throughout the winter and early spring to the southward through Davis' Strait into the Atlantic. It seldom comes in contact with the coast of Greenland below the parallel of Disko, *and there is always an open sea between it and Greenland as far up as Holsteinberg throughout the winter.* The second is the Spitzbergen called also the "store ice," which as has been shown comes down the east coast of Greenland around Cape Farewell, and is carried by the current up the west coast at times even to the Arctic Circle, but by which time it is usually pretty much broken up, and if not entirely dispersed, the last remnants are supposed to return southward by Davis' Strait to the Atlantic—so near these two great ice streams approach that vessels bound to the colonies have in the early spring passed up Davis' Strait with the west ice and the Spitzbergen ice on either hand. But as there are two kinds of oceanic ice, so also are there two distinct classes of icebergs, namely, the bergs from the stupendous glaciers far up the west coast of Greenland, and especially in Melville Bay:—these bergs attain an astonishing magnitude, but like the west ice which they accompany, or outsail, they do not come upon the west coast of Greenland below the same parallel, although in exceptional seasons of violent gales such as the last, they may be blown in upon the land a little more to the southward; and I saw some of these *ice islands* last October aground, upon and near Tallert Bank northward of Frederickshaab. The other icebergs are those which accompany the Spitzbergen ice, and may be said to follow its movements. They are launched from the glaciers far up the east coast of Greenland, and from those in the island of Spitzbergen, and besides being originally far less in their dimensions they are exposed during their long passage southward to the warmer Atlantic winds and heavy swells, and are proportionally reduced before their arrival at Cape Farewell. The bergs from the southern glaciers of Greenland are but small, and need scarcely to be taken into consideration, for as they must come out from the heads of the fiords, they surely would not take the ground in again entering the *channel* of the deepest fiords.

With regard to the flotation of ice, it has been calculated that seven-eighths of a cubical mass of ice will be immersed, but icebergs being very irregular in their formation and having usually very peaked and angular summits, whilst below the water they are smooth, rounded, and most frequently widened out. I think that icebergs are not found that draw more water than the proportion of six feet below to one foot of perpendicular height above the water. Therefore in 150 fathoms of water (the very least found in the entrance to Julianshaab Fiord) an iceberg of an elevation above the water of 150 feet, or having an entire perpendicular height of 1050 feet, will there be suspended above the ground, and such bergs are not to be met with in that place."

Remarks upon the Seasons.

"The finest months in the Faröes are June and July, and in these months only should the cable be laid, and then about the last quarter of the moon, because the tides are greater at the full than at the change, consequently the neap tides and immediately after the last quarter, should be selected, as the currents are then inconsiderable. I have already given my reasons for recommending that the cable be laid from Iceland towards the Faröes, not only on account of the prevailing fogs on the east coast of Iceland, but also from the greater facilities for making the coast of the Faröes, and the opportunity that the comparatively speaking shallow water off the N.W. coast would give of shipping and buoying the cable in the event of a sudden gale of wind occurring at the time of laying it. The finest months upon the east coast of Iceland are also June and July, but I was informed that the weather is clearer earlier in the season, in the months of May and June—I suppose from the alternations of temperature being then less frequent. A few hours, however, of clear weather would always carry a ship beyond these mists, which usually hang only on the land. With reference to Faxe Bay station, the west coast of Iceland is generally free from fogs, and the gulf stream which sets round Cape Reikianess, and appears to keep up a continuous flow around Faxe Bay to the northward, passing out by Snœfellssness, also appears to considerably affect the climatic condition of the west coast. Navigation is open all the year round, and the operation of bringing the cable here can be timed to the opportunities for departing from Greenland. A fine pyramidal beacon has lately been erected on the Skagen, and is of great assistance to navigators entering Faxe Bay from the southward."

Conclusion.

Before concluding, it is proper to state that the voyage was one surrounded with much peril, on account of the succession of gales and the extraordinary quantities of ice found in the Greenland seas; never within the memory of man has there been so much and so long a continuation of ice upon the Greenland coasts as during the past year. In the arduous labours of the voyage Captain Young was most ably assisted by Mr. J. E. Davis, Master in the Royal Navy, who by the kindness of Captain Washington, Hydrographer to the Admiralty of the British Government, was permitted to accompany the expedition and take part in the necessary surveys; and his former well-known services under Sir James Ross in the Antarctic regions, and great experience as a marine surveyor, enabled him to render the most valuable assistance in the especial mission of the "Fox," which is acknowledged by Captain Young in his report in the highest possible terms. During the voyage various specimens of deep interest to the geologist and naturalist were collected; a large number of scientific observations were made, and a detailed meteorological journal was kept, which, together with other valuable information and an extensive collection of photographs made with great zeal by Mr. Woods under very difficult circumstances, have been furnished by Captain Young and Dr. Rae to the promoters of the enterprise, with the hope that they will be found to contribute to the cause of science, as well as to the immediate object for which they were made. Time will not now permit me to give further details of this most interesting voyage; but any members of this Society who may desire to make personal inspection of the charts, meteorological tables, logs, reports, and specimens, will be gladly permitted to do so.

Having thus presented to the Society some of the most

valuable and interesting portion of Captain Young's report, I have only to observe that the result of the recent survey has been to remove from my mind the apprehensions which I previously entertained in common with many others, as to the extent and character of the difficulties to be overcome, in carrying a line of telegraph to America by the northern route.

Prior to the dispatch of the surveying expedition we had no knowledge of the depth of the seas to be crossed, with the exception of the few soundings obtained by Colonel Shaffner in 1859, and our information as to the nature of the shores of Greenland in regard to the requirements for a telegraphic cable was equally small.

These points are of vital consequence to the prospects of the North Atlantic Route, and the survey has placed us in possession of satisfactory particulars respecting them. The soundings taken by Sir Leopold M'Clintock will be a guide in the selection of the most suitable form for the deep-sea lengths of the cables, while the information furnished by Captain Young will direct the construction of the more massive cables to be laid in the inlets of the coast. It is not necessary to determine upon the precise landing places and other points of detail in connection with the enterprise at the present time, but the promoters of the undertaking have received ample encouragement from the survey, and from the testimony of competent and experienced voyagers and sojourners in the countries to which the line is to be carried, to warrant them in proceeding with their labours with renewed vigour and confidence. When they have achieved that success which their perseverance and energy deserve, I am sure they will always gratefully remember that their endeavours at the stage of their operations which is now under discussion would have been very much less productive of good results, but for the patriotic foresight of Lord Palmerston in ordering the "Bulldog" on her late successful service; and for the assistance of Sir Leopold M'Clintock, Captain Young, Dr. Rae, and the Commissioners appointed to accompany the " Fox " by the Danish government, as well as others who took part in the cause, whose patience and devotion to their self-imposed work has been above all praise. Nor can those interested in this important undertaking forget the great assistance which has been rendered to them by the Royal Geographical Society.

SURVEY OF THE LAND SECTIONS,

WITH REMARKS UPON THE CLIMATE, PEOPLE, &c.; BY

DR. JOHN RAE, F.R.G.S., &c.,
COMMANDER OF THE LAND EXPEDITIONS.

The LORD ASHBURTON: I will now call upon Dr. Rae to read his paper.

DR. RAE then read the following, viz.:—

WHEN the promoters of the North Atlantic Telegraph were about to fit out expeditions for the purpose of examining a practicable route, viâ the Færöes, Iceland, Greenland, and Labrador, by which to form a telegraph communication between England and America, I volunteered my services for the land portion of the survey.

In this paper I will give a brief account of my observations, and of information obtained from reliable sources regarding the nature of the lands travelled over, and their suitability for the objects of the telegraph.

THE FÆRÖE ISLES.

After a passage of fourteen days from England in the screw yacht "Fox," we arrived, on the 3rd August, at Thorshaven, the capital of the Færöes. It contains about 900 inhabitants.

On the day following, Colonel Shaffner, Lieutenant Von

Zeilau (Danish Commissioner), and myself, accompanied by two Faröese as guides, commenced a journey over Stromöe, our destination being Haldervig, a village near the northern extremity of that island.

Our course for the first two miles was W.N.W., over the shoulder of a hill (named Klubbin), the height of which about fifty feet below its summit was 1048 feet; we then turned more to the northward until we reached the high land immediately south of Kalbakfiord, 1408 feet above the sea level. The walking round the head of this fiord was fatiguing in consequence of the unfinished state of the path. At the end of five hours we reached the top of the pass overlooking Kallefiord, having an altitude of 1179 feet. When we descended to the valley, we took up our night's quarters at the house of Mr. Dam, a farmer, who gave us a hospitable welcome, and provided us with a good dinner of fish, dried mutton, ham, cheese, butter, milk, cream, and coffee.

Next morning I ascended a hill named Skarling, said to be the highest on Stromöe. Strong squalls of wind, with heavy rain changing into snow as we neared the summit, made the climbing difficult. The barometer indicated a height of 2506 feet. The Colonel had in the mean time travelled along the path a distance of five miles, to the house of J. C. Jacobson, where we joined him. We were again hospitably entertained, and after remaining an hour we resumed our journey.

Our active guides led us by the shortest but the most difficult of two routes, the highest point of which was 1711 feet above the sea. We arrived in the afternoon at Qualvig, a village having 132 inhabitants, where we passed the night.

Next morning we traced back the more level but longer route between Qualvig and Kallefiord. We found its highest point to be 1275 feet at 2¼ miles distance from Qualvig. The hill is not too steep for loaded ponies. We hired some of these excellent little animals for the purpose of testing their qualities. They were strong, sure footed, and carried with ease a man weighing over fifteen stone.

From Qualvig to Haldervig the distance is nine miles, and the path, lies close to the shore all the way. We found the " Fox " at Haldervig.

The formation of the Island of Stromöe is almost wholly basaltic, with an occasional thin stratum of red tuffa. Opals are found in the hills north of Kollefiord.

No difficulties of importance present themselves to the placing of a telegraph line over the route examined, which is about twenty-seven miles in length. At three points of the line some expense would necessarily be incurred in improving the paths, so as to make them more easy for loaded ponies to travel over. These places are, the ascent of the high grounds north and south of Kalbakfiord, the descent to Kollefiord, and the height between Kollefiord and Qualvig.

The inhabitants generally appear to be well educated and religious, and so fully aware of the advantages they would derive from a telegraph being carried through their island, that they would use their best efforts to protect it from injury.

Labour is comparatively cheap, the average day's wages being about 1s. 4d. sterling. Our guides, were well pleased to receive 2s. each per day.

The climate is not well suited for the growth of grain, but small quantities of barley and oats are raised, and a few potatoes, turnips, and other vegetables are cultivated. The live stock of the farmers are sheep, horned cattle, and ponies, sheep being the principal and most valuable productive source. The population of Stromöe is upwards of 2600. The chief exports are wool, woollen goods, eider-down, fish, ponies, and oil. The inland transport is, principally, by packhorses.

Two small bays, the one a short distance to the south of Thorshaven, the other at Haldervig, having been examined by Captain Allen Young, were found well adapted for the landing of a telegraph cable; and the route examined by me overland forms the connection of the projected telegraph.

The sound separating Stromöe from Osteröe offers great facilities for the transport of materials, as it is navigable throughout the whole length, with the exception of about 100 yards near Qualvig, for vessels of ordinary size. The arms of this sound, namely, Kalbakfiord, Kollefiord, and Qualvig Bay, afford good anchorages and approach at three points to within a quarter of a mile of the projected route.

ICELAND.

The "Fox" reached Berufiord, on the east coast of Iceland, on the afternoon of the 12th, and anchored in the harbour of Djupivogr, near the entrance of the fiord. From this place the Land Expedition resumed its labours to travel across the island to Reikiavik. About fourteen horses, and two men, to act as guides and pony-drivers, were required. We had two very zealous auxiliaries in the persons of Mr. Weywadt, the Danish merchant, and Lieut. Von Zeilau, both of whom exerted themselves to procure the necessary assistance and accommodations for the journey. Only eight ponies, exclusive of those of the guides, could be obtained at prices varying from 2*l*. 12*s*. to 5*l*. 10*s*. Hoping to complete our number of ponies on the way, we left Djupivogr on the afternoon of the 15th. With the exception of the guides, our party was the same as that when travelling across Stromöe. Our path ran along the south shore of Berufiord, and was rough and stony.

It was getting late when we reached the head of the fiord, a distance of only nine-and-a-half miles in a straight line; so we proceeded to the pastor's house, which we made our home for the night. This worthy man, Sira Hosias, who had been to Djupivogr, overtook us as we were dismounting at his door and gave us a hearty welcome.

It was difficult to make an early morning's start. Our horse-drivers were active and willing enough, yet we could seldom get away before eight or nine o'clock. A lamb was bought for 2s. 3d. sterling. After taking an observation with the barometer we resumed our journey—and ascended to the tableland west of Berufiord by a series of four steps. The path which is formed among stones, gravel, and earth, might be much improved by a very little labour. Two observations for altitude were obtained. The first about halfway up, giving 891 feet, the last, near the top, 1282 feet, at which the latitude 64° 49′ 3″ N. was also observed.

From this point our path lay nearly due north, for 8 miles, to a small lake 426 feet above the sea level. After travelling 7 miles farther in the same direction we arrived at Thingmuli and took up our quarters in the church, where we found ourselves very comfortable. The clergyman Sira Biarni was a kind and good man. A strong horse suitable for either pack or riding, was bought here for 3l. 8s.

Our course for five miles was north, along the slope of a hill. We then travelled west until crossing the ridge, when we turned to the south-west, and reached in a short time Hallormstadr on the banks of Lagar Fliot. Hitherto we had been surrounded by a dense fog, which we emerged from on descending the hill. We here allowed our horses to feed for an hour, and an observation with the barometer gave the altitude of our position 528 feet, that of the river which was upwards of a quarter of a mile distant being 90 feet lower. This river has its source in the Vatna Jökull, and the muddy and white colour of its water indicate its glacier origin. As far as we could see to the northward the river widens into the form of a narrow lake having little or no current. This was the case we were told for 30 miles or more. As we rode to the southward along the stream we passed through a grove of small birch trees, many of which were from 16 to 18 feet in height. At four miles above Hallormstadr the lake-like river ends, and four miles further up we came to the ferry. The river was here about 170 yards wide, the current strong, and the water so deep from the recent rains that the horses had to swim when crossing. The ferry-boat carried us all over with the baggage at two trips. We walked to the parsonage of Valthiofstadr, 2 miles distant, and met with a most kind reception from Sira Pietra. Taking the direct route to Valthiofstadr, it may be reached in one day from Berufiord, as the distance is about 26 geographical miles, and we learned that the road was not bad.

The morning of the 18th was very beautiful. Our horses having strayed during the night, we were detained some hours. We here bought another horse for 2l. 14s.

After riding eight miles along a fine level path, we turned to the N. W., and commenced the ascent of a steep hill, up which we had not gone half way before we were in a thick fog. Fortunately we had engaged a guide, otherwise it would have been difficult to have kept the proper track.

It was half-past nine, and very dark, when we arrived opposite to Bru, where we had to cross a river on one of those curious swing-bridges, before reaching the house. This conveyance was about 2 feet 6 inches long, 2 feet wide, and 2 feet deep, suspended by pullies to two ropes, which stretch over the river at a height of thirty feet above the stream, which is about seventy feet wide.

Our day's ride had been long and fatiguing, but there were only two parts of the road by any means difficult. The first being the ascent of the hill in the early part of the day already mentioned, and the other where we descended to a small stream, about seven miles distant from Bru. In both instances the ground is of such a nature, that the paths are capable of easy improvement. The heavy rains had made a portion of the road rather swampy.

Sunday, 19th.—Taking with us a guide we started for Mödrudalr. For twelve miles our course was north, we then turned to the westward, which we kept all the way to Mödrudalr, where we arrived at half-past six. The roads were good throughout the day's journey and we passed great quantities of dwarf willow,—at eight miles from Mödrudalr we traversed a perfectly desert plain flat as a bowling green, and covered with black sand and gravel, the débris of lava. Sigurder Jonsson, the owner of the comfortable farm-house gave us a most hearty welcome. The farm is extensive, and produces an excellent crop of grass and quantities of dwarf willow, which, when cut and dried, furnishes excellent fodder for both sheep and cattle. Mr. Jonsson possesses 600 of the former, three or four of the latter, and a number of horses.

Mödrudalr is situated in a beautiful plain extending to a long distance north and south. Far to the south, at least forty-five miles off, you see one of the peaks of Vatna Jökull, having a deep snow filled cleft in the centre. Within fifteen miles to the south-west is Herdubreid, one of the highest mountains in Iceland.

20th.—In company with a Mr. Skulason, who was going in the same direction as ourselves, we left Mödrudalr at nine A.M.

Our course during the whole day's travel was north with a

very little westing. The road was good, and we arrived at Grimstadr, twenty-five miles distant, at half-past two, P.M.

21st.—Mr. Skulason still gave us the advantage of his company. We arrived at the ferry on the Jökulsa Axarfiordr at 9.15 A.M., which is four miles from Grimstadr. The river is 150 yards wide, the water deep, and the current very strong. The horses had to swim about half the way. We crossed in a boat. The banks of the river are of fine black sand. The water was white and muddy, bearing the characteristics of a river that has its source from a jökull, or glacier. Its west bank is 951 feet above the sea-level. When we had ridden sixteen miles from the river, we arrived at an immense bed of very rugged lava in a valley to our left, and in one place there was an appearance as if it had filled up the bed of a river. We rode for seven miles along this lava, and then turned aside to visit a number of boiling mud-springs.

Before reaching Reikialith we had ridden among or close to the most recent lava we had seen. The only object I could compare the rough lava beds to, except in colour, was a field of ice that had been floated at high water to a low flat beach, covered with large boulder rocks, which, when the tide ebbed, broke up the ice into all sorts of forms. Reikialith is situated on the shores of Myvatn (Lake), which is very irregular in form, and studded with rugged lava islands. Our course to-day was nearly west, and the distance travelled fully thirty miles.

22nd.—The first portion of the route was crooked, to avoid holes in the lava which were overgrown with moss and grass. A ride of six and a half hours, including an hour of stoppage to graze horses, brought us to the ford of Arndisarstadr, on the Skialfandafliot. It was about 100 yards wide, two feet deep, the current strong, and the water white and muddy. We reached the house of the worthy pastor (Sira Pallson) at Hals in the evening.

Hals is in lat. 65° 44' N., and was the farthest north point reached by us. I learned that, during the winter, snow occasionally falls to great depth, and is blown into deep drifts. The cold is not usually severe, the lowest temperature being 20° or 21° below zero of Reaumur, equal to 13° or 15° below zero of Fahrenheit, and this occurs but rarely.

Thursday, 23rd.—We resumed our journey, having with us Mr. Eggerd Olafsson, a young student, who most obligingly offered his services as guide. After travelling south two miles along the small river that flows northward past Hals, we forded the stream. We crossed to the west the ridge of hills about 1900 feet high, that lies between Hals and Akreyri. This last-named place is next in size to Reikiavik. As soon as we appeared from under the fog on the hill side, the twelve vessels at anchor in the harbour hoisted their colours. Akreyri is built at the head and on the west shore of Eyjafiord. Its harbour is sheltered by a spit of land that runs half way across the fiord where its width is about a mile. From the beginning of November to the end of March or April, the navigation is usually closed by ice; but during the summer months there is considerable trade at this place. The valley is one extensive and productive grass meadow running southward for nearly thirty miles, on which a great number of persons were occupied haymaking. The river I found to be navigable for boats drawing about two feet of water, to the distance of twenty-five or thirty miles. Its width varies from twenty-five to eighty yards. The path was good and level, fitted for a waggon in summer, or for a sledge in winter.

Early in the evening we came to Saurbær, and were most cordially welcomed by Sira Thorlacius—a clergyman distinguished for his goodness and learning. We here made our arrangements for the next three days' journey, which lay through an uninhabited part of the country. The guide lived at some distance, and our young student rode to his house and engaged him.

24th.—From Holar, the path led us south for four miles along the east bank of the stream, we then commenced an ascent of the high grounds in a south-west direction. The hill was 2868 feet in height, being the greatest altitude we had yet passed over. We reached the top in an hour and a half; although the road was now pretty level, and many of the larger stones had been removed to one side, the path was not good enough to permit us to ride fast.

We made many detours so as to avoid the rougher portions of the ground, but the general direction of our travel was to W.S.W. In the evening a thick fog came on, and as the cairns of stone set up as marks became less frequent, the guide lost his way, and I had to put him right by the compass. About half-past eight we pitched our tent on the bank of a small stream, where there was a little grass for our tired horses. Generally speaking, when travelling over Iceland, sufficient dry willows and willow roots can be picked up for cooking, but here none could be found, so we had recourse to an "Etna" and some alcohol we had carried with us, for the preparation of our coffee.

25th.—Our position by computation was in latitude 65° 8′ N., longitude 18° 53′ W. Height by observation above sea, 2385 feet. The fog still continued this morning, but not very thick. We had to put back a short distance to recover the proper route; we then crossed a small white water stream, very rapid and stony, named Jökulsa Eystri. Our course was generally S.W. by W., marked by little heaps of stones. As usual, we stopped twice to-day to grass the horses, near some lakes, where several swans were seen. We would have made a long journey to-day, but fog again came on in the evening, notwithstanding which our guide insisted that he could find the way. His confidence in himself was misplaced.

Two observations for height were obtained to-day. One at Pollar, latitude 65° 5' N., 19° 2' W., gave altitude 2368 feet; the other, 6 miles S.W. of Pollar, 2463 feet. Both these places afford good grass and water for horses. In latitude 64° 45' we crossed nearly two miles of very rough lava, extremely difficult to travel over. A small amount of labour would remedy this evil. Further on we passed quagmires of very adhesive clay. These by a little care can be easily avoided.

Between the latitudes of 65° and 64° 45' N., and in long. 19° 20' W., we crossed a number of rivers, very rapid and some of them two feet deep, flowing from Hofsjökull, which lay to our left. Many of these streams appear to change their beds, or to spread out to considerable extent during thaws or rainy weather, and could be crossed with difficulty by a telegraph wire, were it not that I noticed that at some points the water was kept within bounds on both sides by solid barriers of rock or lava, generally not more than 20 yards apart.

26th.—Our position by account was in lat. 64° 40' N., long. 19° 33' W., and height above sea 1983 feet. It was ten o'clock before we got away. Our road was better than for the past two days, and we went ahead faster. About half-past three P.M., we arrived at the Hvita River, a mile or two below its source from Hvitarvatn, having come twenty-two miles, down hill, the incline being very gradual. The altitude here was 1580 feet.

The river, which is 120 yards wide, could be forded by the horses, but they were very nearly swimming; so to save our baggage from getting wet, we had a boat brought across, and were ferried over. After permitting the horses to feed for an hour and a half, we travelled westward for five miles round the base of the craggy and lofty Blafell. We now passed over a ridge 400 feet higher than the ferry on the Hvita.

From this point we had a very extensive view of more than

fifty miles down the valley of the Hvita, with its numerous lakes and boiling springs, the clouds of white vapour from the latter indicating their positions. We now made a south course, and at the end of seven miles again came to the banks of the Hvita, along which we found an excellent path, over which we trotted at a great rate until late, when the guide again lost his way, and left us at a good grazing place to search for a house in the neighbourhood.

27th.—After waiting until half-past eleven, in vain, for the return of our truant guide, we unpacked our tent, and lay under it, having no poles to pitch it with. The guide joined us at five in the morning. He said he had wandered about all night in search of his friend's house, which was within less than ten minutes' ride of where we were. We reached Haukadalr at a little before six, and took up our quarters in the church, where we had a good and substantial breakfast.

We then rode forward to the Geysers where we found Lord Milton encamped. We collected specimens, and sounded the Geyser, obtaining seventy-eight feet depth. The barometer gave the height of the position 626 feet above the sea in latitude by observation 64° 18′ 16″ N.

Our breakfast at Haukadalr had not been so superior as to prevent us enjoying the good things that Lord Milton most kindly invited us to join him in partaking of, and after a parting glass of champagne with his lordship, we rode for sixteen miles further on our way to Reikiavik, and then took up our night's quarters at Laugarvatn. Near this house are several boiling springs of very pure water, which are used for cooking.

We arrived at Reikiavik at eight o'clock on the morning of the 29th August, all well, but our horses very much used up by a journey of nearly 450 statute miles.

The results of this journey went to prove that there would

be no serious difficulty in carrying a telegraph across Iceland by the route travelled over. Doubtless considerable expense would be incurred in repairing the old paths, so as to make them more easy for loaded pack-horses, and in making new ones to shorten distances; but this work will be materially facilitated, as the Icelandic Diet has appropriated a considerable amount of money to be paid annually for this purpose. The six largest rivers that we crossed had high well defined banks, that showed no indication of ice action or of changing their position.

A shorter, and in every respects a better route across Iceland for telegraphic purposes than the one described, is that marked on the chart in a dotted line. This route, from Berufiord as far as Mödrudalr, in lat. 65° 17′ N., long. 16° W., is nearly the same as that followed by us. From this point, instead of running northward, it strikes nearly west for 45 miles, over what is said to be not a difficult country, to Isholl, a farm on the Skialfanda river. Following up the west bank of this stream to near its source, you cross the centre of Iceland in a south-westerly direction, by the Sprengisandr road, until you fall upon the head waters of the Thorsa. Trace this stream to south-west, keeping on its left bank, to avoid the numerous jokul streams that enter it on the right, until reaching lat. 64° 20′, where the river would be crossed. The course then would be west to the Hvita and the Geysers. On nearly fifty miles of this route there is little or no grass, but depôts of hay can be established. Having measured on the charts four different routes from Berufiord to Reikiavik, the distances are about as follows:—

	Geographical Miles.
Route travelled over, cutting off several unnecessary detours	310
By contemplated telegraph route, *via* Sprengisandr . .	250
In a straight line, keeping north of Vatna Jökull . .	210
Along south shore of Iceland	260

SURVEY OF THE LAND SECTIONS.

The modes of transport through Iceland are by pack-horses, waggons, and in winter on sledges. Of these the pack-horse is by far the most general. These little animals are remarkably sure-footed, and so strong that they can carry a load of 200 or 250 lbs. with apparent ease. They are easily kept in condition with no other food than grass or hay. Their prices vary from 2*l.* to 3*l.*, those for riding being more expensive. The pack-saddle in general use is an extremely primitive affair, the pads employed to protect the back from being injured being composed of turf which has been well dried, and a portion of the mould beat out of it. Boats might be used with advantage on some of the rivers.

The population of Iceland amounts at present to some 60,000; at one time it is said to have been as high as 100,000, but the ravages of epidemic diseases and other causes reduced the numbers to less than those at present on the island.

The masses of the people are able and active, harmless and honest. Wherever we went, we were received with much kindness and hospitality, and even at the poorest cottages, milk, coffee, and brandy were handed to us. All classes seem more or less educated, and the Lutheran religion prevails. The chief occupations of the people are fishing and farming, both being combined when the farms are near the sea. The women spin, knit woollens, and weave cloth for home consumption. The farm live stock consists of sheep, ponies, and horned cattle; the two last are of small size. Of these, the sheep are the chief source of wealth. A farmer having 800 or 1000 sheep is considered wealthy. The usual food is mutton, fish (fresh and dried), rye-bread, butter, cheese, milk, one preparation of which, named *skuer*, is much used.

The price of labour varies from 1*s.* 2*d.* to 2*s.* 8*d.* per day, according to the season. During the haymaking, in the months of July, August, and September, it is highest.

Reikiavik, the capital of Iceland (a town of 1500 inhabitants), has been so often described by others that it is needless for me to say anything on the subject. A little thin ice forms along shore near Reikiavik during calm weather in the winter time, but the first breeze of wind disperses it.

GREENLAND.

The "Fox" sailed from Reikiavik on the 31st August, for Greenland. On the 2nd of October we reached Fredrickshaab.

There is a Danish superintendent, a clergyman, and several clerks at this place, and about two hundred Esquimaux. These Esquimaux are civilised;—sober, honest, and faithful, apt and willing to be instructed; attentive to their religious observances, and thankful for kindness. The evening amusement was dancing. The principal food of the natives is fish, seal, whale, a few ptarmigan, waterfowl, including eider duck, with biscuit and coffee, imported from Denmark. Large quantities of a small fish (the kepling), called by the Esquimaux "amaset," are caught in scoop nets in the summer, and dried on the rocks. These are laid up for winter food, and sometimes given to the cattle.

After lying here eighteen days, the "Fox" sailed on the 20th of October for Julianshaab, at which place she anchored on the evening of the 22d. Julianshaab is one of the principal stations on the coast.

On the 24th I learnt that it had been decided to sound and examine the Fiord of Igalikko, which ran by Julianshaab. During the time that the "Fox" would be employed on this service, which I was told by Captain Young might probably occupy four days, I thought with Colonel Shaffner that a short journey should be made to the interior of the country, for the purpose of ascertaining the practicability of travelling over it. The use of one seaman and a whale-boat were obtained from Captain Young, to enable us to return from the head of the fiord to Julianshaab. Four Esquimaux women were engaged as rowers. At sixteen miles inland from the fiord a heavy fall of snow stopped further travel. After an absence of four days, we returned to our boat, but found that the fall of snow, followed by unusually cold weather, had already caused the fiord to freeze up for many miles. We had enough of provisions, and were supplied with some excellent fresh mutton, milk, and butter, by an Esquimaux that lived in the neighbourhood, to whose house we removed. The frost continued for several days with unusual severity, and made the ice strong enough to enable Captain Young (after coming half-way up the fiord in a boat) to travel over the ice with a sledge party from the "Fox," to our relief. Another party of men, sent by Superintendent Möller from Julianshaab to aid us arrived at the same time. We all returned next day (the 6th) to the "Fox."

At Julianshaab, as at Fredrickshaab, nothing could exceed the kindness and hospitality of the resident Danish gentlemen. Mr. Möller, the Superintendent, Mr. Höyer, his Assistant, the Doctor, and others vied with each other in paying us attention.

The chief exports of the place are whale and seal oil, fox skins (blue and white), bear skins, and eider down. A few cattle, goats and sheep are kept. The hay is usually collected at the summer encamping places of the natives, and must be very nutritious, as I was informed that one small cow during the past summer, had not only yielded sufficient milk and cream to supply the family, but also to make eighty Danish pounds of butter. The natives here, as at Fredrickshaab, are honest, docile, and well conducted, doing great credit to the Danish Government. The prevailing form of worship is the Lutheran.

The result of this expedition, as far as regards the land portion of it over the Færöe Isles and Iceland, was extremely favourable to the practicability of laying down or erecting a telegraphic wire. The question in Iceland will be, whether the telegraph should be carried across the whole island from Berufiord to Faxe Bay, or only from Portland Bay to the latter place. The latter will reduce the distance on land from about 250 to 90 miles.

THE FIORDS OF SOUTH GREENLAND,

AS TO THEIR DEPTH, MUD BOTTOMS, &c.; BY

J. W. TAYLER, ESQ., F.G.S.,

LATE A RESIDENT AND MINERALOGIST IN SOUTH GREENLAND.

—⁜—

The LORD ASHBURTON: I will now call upon Mr. J. W. Tayler to read his paper.

Mr. TAYLER then read the following, viz. :—

I HAVE been desired by the promoters of the North Atlantic Telegraph to give you a short description of the Fiords of South Greenland, and to point out the advantages they offer to the landing of the cable on that coast.

Any one who has travelled in Norway will have an idea of the leading characteristics of fiords in general, but as many of the Society have probably not had that opportunity, I will endeavour to describe, as well as I can, the general appearances of fiords in Greenland.

The land of Greenland is very elevated, the average height of its mountains being not less, perhaps, than 1500 feet, and in some places exceeding 6000 feet, above the level of the sea. It appears that at the time of the elevation of the west coast of Greenland, a chain of mountains of about 50 miles in breadth, running about north and south, was acted on in a wave-like manner, *i. e.* leaving depressions nearly equal to the elevations and more or less at right angles with the direction of the chain. These depressions or long valleys into which the sea runs constitute the fiords:—they vary in breadth from one to eight miles, and run up into the interior from ten to sixty or more miles.

The scenery in these fiords is magnificent, perhaps unequalled. The lofty mountains—rugged, precipitous, and barren—with patches of ice, (projections from the great interior glaciers), and snow unmelted by the summer's sun; with valleys half filled up by enormous angular blocks of stone detached from the sides of the steep mountains by the alternate frost and thaws; the solitude, and the almost total absence of life, animal and vegetable, make up a picture of indescribable desolation.

In other places the more rounded and sloping mountains are covered with green and yellow moss. Grass, heath, and wild flowers grow in the valleys; whilst in some still more favoured and sheltered dales, miniature forests of Arctic willow six feet in height are pointed out by the Esquimaux as proofs of the extraordinary excellence of the climate of Greenland and fertility of the soil. Of the latter material there is not, however, in many places sufficient to bury the dead, and they are compelled to place the body on the surface and form a grave by building up stones around it.

These fiords with grassy dales offer the most pleasant places of abode in Greenland; in fact, it is only to the more fertile parts of the fiords that the name of Greenland is at all appropriate. But these parts of the fiords form a striking contrast to the outer coast of Greenland.

The Danish settlements are mostly at the entrance of the fiords, for the convenience of seal hunting and of shipping; but the old Scandinavians who settled in Greenland in the ninth century brought cattle with them, and therefore established themselves at the interior ends of the fiords and bays, where grass was to be found. The ruins of their habitations, constructed of very large blocks of stone, are still to be seen at all the more fertile places. Judging from the number of ruins and the accounts in the Icelandic Sagas, their number must at one time have reached about 10,000. But the Scandinavians of Greenland have perished; cattle no longer graze in the valleys as then; and some heaps of stones are all that remain to show the enterprise of those early western pioneers.

The Icelandic Sagas contain descriptions of most of the fiords of South Greenland, and of the chief settlers in them. Perhaps the most notable in this respect is the Fiord of Igalikko. This fiord ends in two forks or arms;—in the northern stand the ruins of Brattelid, the first town in Greenland, built by the first settler, Eric the Red, in 986; in the other was built the town of Garde, the residence of the Bishop of Greenland. These two towns vied with each other in the claim for precedence—Brattelid claiming it on the ground of its being the first erected, and the residence of Eric the Red and his descendants; whilst Garde asserted its superior worthiness in being the residence of the bishop. After much wordy quarreling and sundry duels, Garde appears to have triumphed, and was henceforth considered as the capital of Greenland.

But Garde now shares the ruin and desolation of Brattelid, with nothing else to recommend it to our further notice. Not so however with its rival Brattelid. In the time of Eric the Red, Anno Domini 1000, there sailed from the fiord of Igalikko and from the town of Brattelid an expedition of discovery. These enterprising Scandinavians were not contented with having discovered the vast territory of Greenland; this appears to have only stimulated their thirst for further discoveries, and it may perhaps also be that after two or three years' residence in Greenland, they found it was not the El

Dorado they had dreamed of; however, the expedition sailed to the west and south, and finally discovered the continent of America. They found good prairie land which they called Markland, and sailing on they came to land bearing wild grapes in great abundance, this they called Viinland. They wintered in America, left some settlers (who after some time succumbed to the natives), carved their runics on the rocks, and taking in cargoes of timber and supplies of wild grapes sailed the next summer and safely returned to Brattelid.

Thus the credit of discovering America is certainly due to the Scandinavians of Greenland. And Columbus when he visited Iceland some years previous to his celebrated voyage, no doubt read the accounts of the discovery in the Icelandic Sagas. But as these accounts were suffered to remain almost unknown to the world, and as the navigators of northern Europe had in his time forgotten the route discovered by their more energetic forefathers not only to America but also to Greenland, (which was then called the "Lost Land,") all honor is certainly due to Columbus for the re-discovery of America.

When Eric the Red first settled in the fiord of Igalikko, the Christian religion had not reached Iceland or Norway. Thor with his hammer reigned supreme. But Thor and his worshippers have sunk into oblivion, bequeathing us his hammer, the symbol of industry. The fiords of Greenland are about to witness a new era of enterprise and engineering skill, ere long the North Atlantic Telegraph Cable will repose at the bottom of one of its fiords, again uniting with Europe the countries of Iceland, Greenland, and America.

In the neighbourhood of high land the water is generally of great depth, and the fiords of Greenland form no exception to the rule; in fact, it is only near the lower and more sloping land that vessels can lay at anchor, and this only in small coves so near to the land that it is the general practice to make fast to the shore by hawsers. Outside of this the water deepens rapidly, and in the middle of many of the fiords there is some 500 fathoms water; I should say that of the deeper fiords 300 fathoms will be found to be the average depth. In some places where the rocks are nearly perpendicular, a fishing-line of 100 fathoms fails to reach the bottom within a few yards of the shore.

Looking at the map of Julianshaab's district, you will see that some of the fiords terminate at the continental ice, whilst others do not reach it; the former fiords have glaciers, the latter have not.

These glaciers are the outlets of the continental ice, which has a motion from the interior towards the sea coasts, and as the deep valleys or fiords are the only outlets, the ice is forced into them until, by projecting from the land into the sea, or fiord, portions give way and break off, owing to not being sufficiently supported by being adequately immersed, or owing to the rifts and chasms which exist in the glacier. The larger of these portions, when thus detached from the glacier, constitute icebergs.

The glaciers in the fiords of the southern parts of the west coast of Greenland are not very large, and consequently their icebergs are never of great dimensions. I do not think any iceberg produced by the glaciers of the fiords within 100 miles north or south of Julianshaab would ground in 60 fathoms.

The glaciers bring down with them boulders, sand and much fine clay, the result of attrition; the boulders are always rounded, owing to the severe abrasion they have undergone by being transported over the rocks below, whilst under the enormous pressure of the vast thickness of continental ice. This glacial clay floats suspended in the water, several miles from the glacier, rendering it turbid or milky, and depositing

F

itself gradually throughout the whole length of the fiord. So vast has been the quantity, and so long the period of time during which this transport of clay has proceeded, that some fiords have been so completely filled up by it that they are only navigable in boats of light draught and at high water.

From this choking up of the fiord the glacier has ended in being unable to longer launch its icebergs; it has therefore found a new outlet through some other valley, where it will repeat the process of gradually filling up the fiord.

As nearly all the fiords have, or have had, glaciers in them bringing down the clayey deposits I have mentioned, the bottom must be of soft material. If the cable be taken into a fiord having a glacier, I think the clay which will be gradually deposited over it will be of great service in protecting it from injury by marine animals or other damaging agents.

The existence of these fiords is extremely advantageous to the carrying of the telegraph cable to Greenland, and there bringing it on shore. Were it not for these, some difficulties might have been met with in finding a suitable place for the landing, owing to the ice streams on the outside coast; but in several of these fiords, Tessermiut, for example, the water is of such depth as to preclude the possibility of icebergs grounding upon the cable, and the almost perpendicular mountains forming parts of the lateral coasts of the fiords, and the deep water at their bases, offer excellent situations for leading up the cable from the middle of the fiord to the shore, without exposing it in the slightest degree to the grounding of icebergs upon it.

In conclusion, I beg to state, that from the results of seven years' observation in Greenland, I am of opinion that neither the ice nor the configuration of the coast, will offer any impediment to the successful laying and landing of the telegraph cable in Greenland.

THE ELECTRIC CIRCUITS

OF THE PROPOSED NORTH ATLANTIC TELEGRAPH; BY

COLONEL TAL. P. SHAFFNER, LL.D., F.R.G.S.,

OF THE UNITED STATES OF AMERICA.

—◆—

The LORD ASHBURTON.—We have now come to the end of these papers which more particularly come within our province as geographers. There seems to me to be something wanting still. Physical geography is only interesting in so far as the results detailed are associated in some manner either with human feelings, human sympathies, or human interests. Now, the great purpose which has given attraction to the details which we have just listened to, is that of ascertaining whether the physical conditions of these northern regions are suited for the carrying out of the great enterprise which both Continents have at heart. Therefore, it seems to me you will hear with great interest how these details dovetail, as it were, into the necessities required by the establishment of the electric telegraph. The particular features of this scheme will be put before you by Colonel Shaffner, and I trust you will take the same interest in it that I am sure I myself shall.

COLONEL SHAFFNER then read the following, viz.:—

As preliminary, I will briefly refer to the respective circuits proposed on the north-about route, as marked upon the large map before the Society.

LANDING PLACES AND LENGTHS OF CIRCUITS.

SCOTLAND TO FÆRÖE ISLES.—One end of the cable for this section will be landed in one of the many safe bays in north Scotland—the precise place has not been determined. The other end it is proposed to land in a beautiful bay near Thorshaven. This section will be some 225 miles, and the depth of the sea not exceeding 254 fathoms; bottom, mud and shells.

FÆRÖES TO ICELAND.—One end of the cable will probably be landed at Haldervig, near the north of Stromöe Isle. Captain Young strongly recommends this bay, and that the other end be landed in Berufiord, Iceland, a very good place, with deep water and muddy bottom. The cable to this place will be about 240 miles. Depth of sea, maximum, 683 fathoms, bottom mud and shells. It may be found more advantageous, for reasons not necessary now to be discussed, to carry the cable more westward, to or near Portland, and to which place it can be laid on a muddy and sandy bottom, in water of good depth. Both of these places, namely, Berufiord and Portland, are free from any volcanic influences whatever, and ever have been, as far back as the discovery of that remarkable island in the year 863.

ICELAND TO GREENLAND.—This section will be the longest of the series, between 600 and 700 miles. Captain Young has reported in favour of Hvalfiord, a little north of Reikiavik, in the Faxe Bay, but in order to economise as to length of cable, it is quite probable that a more westerly place will be selected, on the south side of Faxe Bay. The other end of the cable he recommends be landed in Julianshaab fiord, on the south-west coast of Greenland. He examined that beautiful bay, and found it to contain deep water, with muddy bottom; and he states that it is his decided opinion that bergs cannot reach the cable when laid in it. The reports of others concur in this opinion. There are other fiords near Cape Farewell, equally favourable. Tessermiut and Illoa fiords are considered well suited for the cables.

Heretofore it was contemplated to land this section of the cable upon the east coast of Greenland, south of latitude 61° north, in or near Prince Christian's sound, and then, either to carry the cable out at the other end of the sound, or to connect this section with the next by a cable around Cape Farewell; or, by a line across the land, avoiding the inland ice. For the present, the intention to land on the east coast has been abandoned, not because it has been found to be impracticable, but because it has not been proved to be practicable. It is now proposed to carry the cable from Iceland, around Cape Farewell, to one of the fiords on the west side, and from the same fiord run the cable to Labrador. Hereafter the Company may find it best to land on the east coast, and carry out the original intentions as above stated.

GREENLAND TO LABRADOR.—The cable will start from Greenland, and land at or near Hamilton's Inlet. The soundings taken by Sir Leopold M'Clintock show 180 fathoms interior from the outer rocks on the coast, so that the cable can be laid into the inlet from the sea in water sufficiently deep to place the cable beyond the reach of icebergs. If, however, the depths be found more favourable from the sea into Byron's Bay, near and south of Cape Harrison, that place will be equally satisfactory, all questions being considered; and from thence the cable can be carried into Hamilton's Inlet, through one of the several channels connecting Byron's Bay with the Inlet. Length of section about 510 miles.

Before the respective cables are laid, each of the places will be carefully sounded, and buoys will be placed indicating the deep trenches; and, besides, steam tenders will be in readiness to serve as pilots. Every precaution will be taken to obtain the most complete information, in regard to the depths of the bays and of the seas.

THE ELECTRIC CIRCUITS.

I will now explain to the Society the respective electric circuits, as practically applied to commercial telegraphy. I propose to illustrate the mode of manipulating one or more circuits singly and combined: when I say singly, I mean from one station to another, and the employment of one voltaic series: when I say combined, I mean the coupling together two or more circuits in succession, all of which are brought into action by the manipulation of any given one. The diagram, I. represents

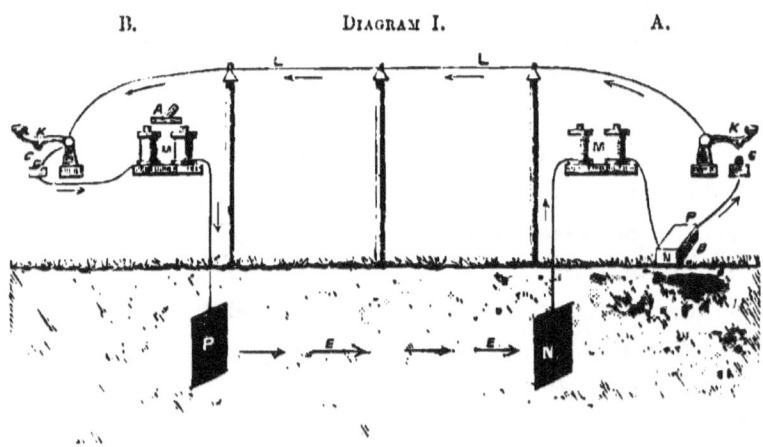

a voltaic current traversing a line on poles: M. is the magnet, B. the battery, and K. the key, or circuit closer at contact point C. of station A., A. the armature of the lever seen in the apparatus at station B. By this diagram it will be seen that the course of the electric current leaves the positive end of the battery B., and passes through the metallic key K. when brought into contact with the metallic point c., thence with the line wire to station B., and there through the respective instruments to the metallic plate P., thence through the earth E. E. to the metallic plate N., up with the wire to the magnet M., thence to the negative pole of the battery B., which completes the voltaic circuit. Before proceeding further, it is proper to say that the course of the current, when it takes the earth at station B., remains a mystery in physics, and I assume its return to the battery from whence it originated. By elevating key K. the circuit will be broken, and it is the opening and closing of the circuit, in time methodically arranged, that constitutes the alphabet for the transmission of intelligence.

I will next call the attention of the society to diagram II., which illustrates the respective circuits contemplated to be organised on the North Atlantic Telegraph. The first circuit connects Scotland with the Færöes by submarine cable of some 225 miles in length. The course of the voltaic current runs through the cable, starting from Scotland, passing through the apparatus at the Færöes, descending into the earth, and thence direct to the source of its origination in Scotland. The current of this circuit passes through the spools of wire at the Færöes constructed as the one which I hold in my hand, and when the wires are thus charged, the cores, or iron centres, become magnetised, and they attract down the armature, which causes the other end of the lever to form a metallic contact so as to close the succeeding circuit extending to Iceland. The course of the current on this second section extends from the Færöes to Iceland, thence throguh the apparatus, similar to that at the Færöes, and then returns to the battery, from whence it sprung. In like manner the circuit from Iceland to Greenland is opened and closed, and the same action takes place on the circuit from Greenland to America. It will be observed that these circuits are not of equal lengths, but the science of telegraphy has been so carefully studied by the practical manipulator, that the conductors can be adapted to the requirements of each circuit, if desired.

The limited time allowed at this meeting of the Society prevents me from going further into detail in explanation of the sciences and the arts pertaining to the subject under consideration. But in order that the Society may more clearly understand that which I have already said, I have arranged the models of apparatuses placed in different parts of this room. Before me is an apparatus supposed to be at North Scotland, that on my right the Færöes station. I will now call the Færöe station by the opening and the closing of the circuit as shown in diagram II., section 1, and it will be seen that the bell at Færöes is made to ring by the instrument at Scotland. I will next call Iceland, which is the station immediately beyond the Færöes, as seen in diagram II. When I call Iceland, the Færöe station connects or couples on the circuit to Iceland; when Iceland hears the call by the ringing of his bell, he immediately responds. The next station desired is Greenland, and when that station is called, Iceland couples on the circuit to Greenland, and in like manner Greenland couples on the circuit to America, and now I communicate by the combination of the respective circuits from Scotland to America by the simple motion of my finger.

I have reduced this illustration to the most simple form, so as to bring it within the comprehension of the most uninitiated. It has not been done for the expert telegrapher. The apparatuses required at the respective stations to effect the working in the manner described, will be more complicated, and

ELECTRIC CIRCUITS COMBINED.

Diagram II.

DESCRIPTION OF THE APPARATUSES.—At *Scotland*:—K is the key; c, point to close circuit; B, the battery connected with earth plate N from negative pole N. At *Færöe Isles*:—M is the magnet; A, the armature of the lever L; F, the fulcrum post; C, the contact post; B, the battery, with its positive pole P; N and P are earth plates. The black arrows indicate the course of the voltaic current through the cable; dotted line E the current returning to the earth plate N, and to the battery at N. At *Iceland*:—The apparatus is the same as that at the Færöe Isles.

MANIPULATION OF THE APPARATUSES.—When the key K at Scotland is pressed to make a contact at c, battery B sends its electric current in the direction indicated by the arrows to C; through the key K, the submarine cable to the magnet M, through the coils of the magnet M to earth plate P, and thence to earth plate N, as indicated by the dotted line E. In this manner the circuit is completed.

At the Færöe Isles, when the current traverses the magnet coils M, the armature A is attracted down, and the lever L makes a contact with C, which closes the next circuit. The current passes from battery B through the cable to the magnet M at Iceland; then to the earth plate P; then with the dotted line EE, to plate N; then up the fulcrum post F, through the metallic lever L, and the contact post C to the battery B. When the current traverses the coils of magnet M at Iceland, the armature of lever L is attracted down, and a connection is made at C, which closes the next circuit to Greenland. At Greenland there will be a like apparatus, which will close the circuit to Labrador; and at Labrador will be another apparatus which will close another circuit to Quebec. At each place there can be a recording instrument, and despatches can be received simultaneously at the Færöe Isles, Iceland, Greenland, Labrador, and Quebec.

a description of them would occupy more time than is allotted to me on the present occasion. I have upon the table one of the instruments used on the continental lines for the uniting of two or more circuits, by which the end heretofore described is effected.

The respective circuits of the North Atlantic Telegraph will be within the bounds of known practicabilities, and we have no reason to expect other than the fullest success and the attainment of a calculated celerity. There are no air lines or lines constructed on poles that work in circuits more than perhaps 500 miles. There is no retardation on such lines. The telegraph from London to St. Petersburg is about 1900 miles, and upon that range there are eight relay stations. From London to Odessa the telegraph is about 3500 miles, and there are about fourteen relay stations. From London to Constantinople it is about 3200 miles, and there are about twelve relay stations. It is a common occurrence to work to the stations mentioned through the aid of mechanical contrivances. It is in this manner that telegraphic manipulation overcomes long distances, and although the original electric spark will not "leap with one bound" from continent to continent by the North Atlantic Telegraph, the intelligence will pass between the hemispheres with a celerity commensurate with the grandeur of the enterprise and the wants of the age.

The LORD ASHBURTON: The Society has heard, I have no doubt with much pleasure, the very interesting papers that have been read this evening, and to afford an opportunity for a discussion upon them, their consideration will be continued at the next meeting.

The Society then adjourned.

THE ROYAL GEOGRAPHICAL SOCIETY.

MEETING, FEBRUARY 11TH, 1861.

THE LORD ASHBURTON IN THE CHAIR.

A crowded Meeting of the Members and their Friends was held at Burlington House, on Monday Evening, February 11*th,* 1861, *when the Adjourned Discussion took place.*

The President: I need not read over to you again the names of the very interesting papers that were submitted to you on the last occasion of our meeting. We have now amongst us travellers who have gained experience in every quarter of the globe;—many learned geographers, who have been themselves present in those portions of the earth which have been alluded to in these papers: and I trust that we may have their assistance in correcting what they may see to be wrong; in amplifying and illustrating what we have heard. The subjects which come before us, within our province, are the geographical features; we are not engineers, nor electricians, to judge of the power of electric currents to pass from this or that part of the earth; we are not members of Parliament, who are to determine what aid, what facilities by law are to be afforded to men who have devoted themselves to this great enterprise; nor are we capable of judging of the respective merits of the many plans which have been put forward before the public. All that we can judge of are the geographical facts which have been laid before us, and of those facts there cannot be better judges than the assembly that I see before me. (Applause.) It is, therefore, with hopes of receiving instruction and much knowledge, that I call upon those gentlemen, more particularly, who have been in these scenes, to give us the advantage of their experience, in either amplifying, correcting, or illustrating, the several facts which have been brought before us. (Applause.)

Sir Edward Belcher, R.N.: I took a very deep interest in the deputation to Lord Palmerston to get the expedition sent out for this northern telegraph; and I am very happy indeed to find that nearly all the arguments that were then used have been so thoroughly realised. I thought, in the first instance, as has been proved, that we should find a great connecting bank between Scotland and the Færöes, and between Iceland and Greenland. As to the difficulties which were raised with reference to the Labrador shore, I am very glad to find that the reefs, which it was said would entirely prevent any cable being laid across there, have vanished with the ice; they have gone southerly somewhere or other, and have not yet been found; at least those reefs that are there are rather helps than otherwise by preventing the ice coming down upon the entrance. I should like very much to have been present at the last meeting to have understood the nature of the soundings;—and while I am speaking of the soundings, I am sorry to say that before another institution—the Civil Engineers—it has been assumed that we surveyors know nothing at all about soundings—that we do not know our business, and that really some civilians will have to teach us our work. Now, as far as I am myself concerned, during the whole of my servitude I adopted a particular means for getting up the bottom in large quantities, and in such large quantities that I thought nothing of bringing up a hatful at a time. I have six cases of microscopic shells, a very handsome collection, all of which I have got up by that means. The very simple mode that I adopted to sink my lead and to get up the soundings from the bottom was this. The sounding lead was encased by a cylinder, and the lead being conical, the cylinder on striking the ground flew up, and the lower part of the lead stuck to the bottom, picked up its quantity of soundings, and as the lead was withdrawn this cylinder (pointing) slipped down and completely protected whatever adhered to the lead from being washed away. The other one which I used for obtaining shells was a small dredge attached to the lead, and as the lead struck the bottom, the flanges scraped up a quantity of earth.

The Labrador coast takes its name "Labrador" from people having found there golden pyrites, as they termed them in those days, when they thought that they had made a discovery of a golden coast. It is a golden coast, inasmuch as the copper contains a great quantity of gold. In approaching that coast, I think that it would have been as well if the "Bulldog" had been furnished with a fine wire chain, such as our jack chains, as used in sweeping up anchors, and had it connected with an electrical machine. As she ran over the ground she might easily have detected at the instant whether there was any copper there. And as to any chance of the cable being corroded or cut, on approaching the shores of Labrador, or any of the places, I think that if application be made to the civil engineers, there are men of talent who would inform them instantly how they were to obviate it.

And if the cable is to be properly laid down, the ground must be properly examined; for it is quite absurd to attempt to lay cables in the teeth of nature. If that cable is to be connected with the Labrador coast, and that coast runs off in ridges under water, as is very natural, a further survey must be made to find where the hollows run. A correct and a very accurate survey should be made, so that the cable may be laid in the hollows.

I am very happy indeed that the experiment has been made, and that the officers have been so successful in planning the ground; and that we have a chance of connecting our coasts with those of our cousins across the Atlantic. (Cheers.)

Mr. PLINY MILES.—I do not know, My Lord, that I can throw any particular light on the subject of the explorations in Iceland and Greenland. I can, however, mention one or two facts that I noticed while travelling in Iceland a few years since. One great consideration of course will be that of climate; we can ascertain from certain circumstances that that country is not so cold as we should imagine it to be from its name and locality. The small hardy race of horses there subsist entirely out of doors in the winter, without any kind of food being given them in the shape of hay, excepting only those horses which they use for domestic purposes, during those months. The rest are without any kind of food, except what they gather from the open country. I saw in one case a mahogany tree on the south coast, not far from Portland, that had unquestionably floated with the gulf stream from the Gulf of Mexico itself. I was told also by Professor Johnson, that it was not unfrequently the case that spruce or other logs from Norway were found driven on to the northern coast, at the same time that logs of wood floated from the southern portion of the North American continent. They are there found together, the one current having come from Spitzbergen, and the other from the gulf stream. Standing on any mountain or height in Iceland, you will see a green and beautiful landscape; in fact, the natives say, "Iceland is the finest country the sun shines upon," though at some seasons of the year they have very little of the sun to shine. I am satisfied of the feasibility of running the wires from this country to the Færöes and Iceland.

With regard to the absence or presence of trees in Iceland, that has nothing to do with the coldness or the severity of the climate. It is well known that there are trees of large size on the coast of Norway, where the climate is far colder than in any, the most severe seasons that have ever been known in Iceland, or in regions still farther north. It is a fact, that on some portions of the coast of Scotland, and in all parts of the Orkney and Shetland and Færöc Islands, you will not find a single tree six feet high, while they are found in Iceland 18 feet high. I will not attempt to explain, or to give a reason why, we find trees growing in one very severe climate, and not in others, but it is undoubtedly owing, to a certain extent, to the direction of the wind, and to some particular effect of the sea air. At any rate we find no trees whatever on some of our more southern coasts, while, farther north—on the Labrador coast, there are large trees, spruce, pine, larch, and others. So that the fact that there are no trees in Iceland, must not lead persons to suppose that the cold is necessarily extremely severe.

I would just refer to a statement made by a gentleman who read a paper here two weeks ago, with regard to the thickness of ice at Reikiavik. I have a distinct recollection that the lake seldom freezes to the depth of 18 inches. I have been told that sometimes it is not frozen more than two inches during the whole course of the winter, showing most conclusively the mildness of the climate. These circumstances will throw considerable light on the peculiarities of that climate; and prove that it is far more genial than you would naturally suppose from its name and locality. (Applause.)

PROFILES OF THE DEEP SEAS
ON THE
NORTH ATLANTIC TELEGRAPH ROUTE.

[Referred to in the Remarks of Captain Osborn.]

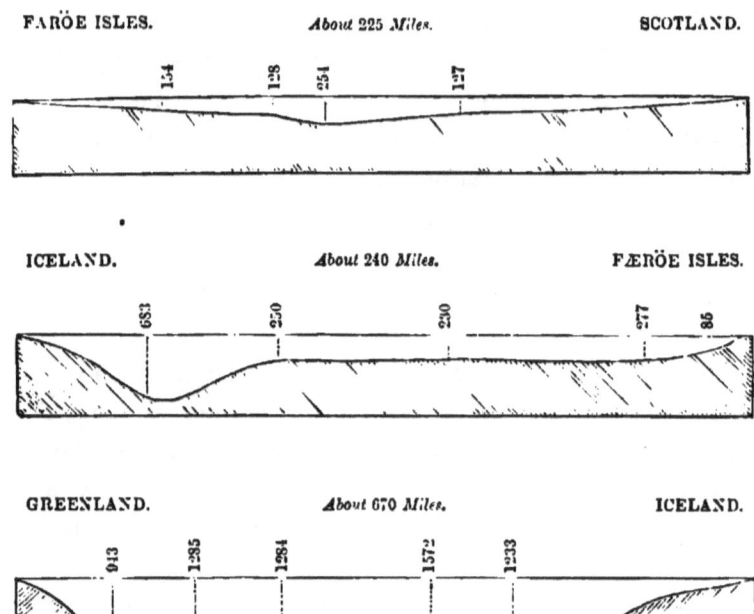

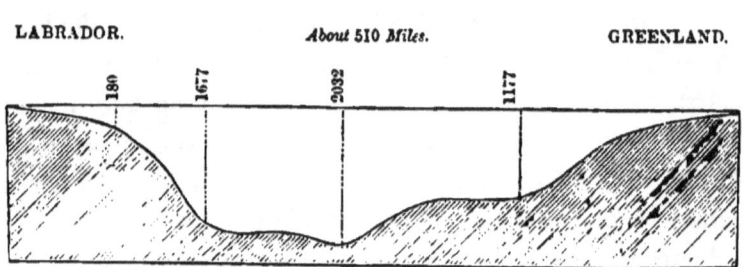

CAPTAIN SHERARD OSBORN, R. N.: I will not attempt to say anything as to the practicability of the route, inasmuch as the distinguished travellers and navigators, who read their papers at the last meeting, have given us sufficient authority to say that there were obstacles on the route undoubtedly, but that all those obstacles are surmountable. Touching the soundings, I would call your attention to the agreeable fact that the entire line of soundings show that there is no depth greater than 1000 fathoms between Scotland and longitude 30 west on the proposed route, or exactly halfway across the Atlantic Ocean. There are then two valleys of deep water, of small width, on either side of Greenland. The soundings diminished, as you see, very abruptly as Greenland is approached, and shallow water will be found probably round Cape Farewell. With regard to the Labrador coast, the principal difficulty was in carrying the cable within the 150 fathoms of water, so that it might not be exposed to the action of the icebergs. I am sure that Arctic men all agree that 150 fathoms would be about the maximum draught of any iceberg, so that the chief object to be secured in placing the cable is to push the 150 fathom mark as near into the coast as possible. They would remember that Sir Leopold M'Clintock had said that he would like to have a more accurate survey of the Labrador coast, so as to carry the deep water a little closer. If they looked at the map they would see 1190 fathoms carried just to the north of Hamilton Inlet, and there was every reason to believe that the bank of Newfoundland—they might call it so, although it extended a long way up the Labrador coast— deepened suddenly there. As far as one might speculate upon the question, I have no doubt that a more accurate survey would lead to the obtaining of such information as would give us a channel leading close into the Labrador shore, near Cape Harrison. I would propose, that the cable there should open out and form, say, half-a-dozen strands running into the shore at certain distances from each other, so that if a berg drifted down and happened to pick up one, it would not pick up all the rest, and by that means the cable in the deep water would be at all times safe and recoverable by following out those strands. There was one feature that struck me particularly, and on which I wrote to Captain M'Clintock. No man would dispute that bergs of ice sounding the bottom would rip it up, just as a plough would a ploughed field; and that if that bank was thus ripped up by the floes of ice, there would be hardly any animal or vegetable life upon it; it would be a kind of subterranean desert. Sir Leopold M'Clintock was continually dredging there, and in a note just received, that distinguished navigator said, "The bank of Hamilton Inlet has from 100 to 200 fathoms upon it. I am of opinion that icebergs which ever drift down there cannot possibly reach the bottom. They ground near the islands at the entrance and are not bergs of the largest size. Shells and other small things were brought up by the sounding machines in 40 or 50 fathoms, *as in other parts of the sea* where icebergs cannot possibly interfere with them." Near Frederickshaab Captain Sir L. M'Clintock says they dredged up in 26 fathoms delicate corals and creatures which could not live at the bottom of the sea if much disturbed by icebergs,—and adds "*so much for the destructive propensities of icebergs.*" How all this came about was a question I do not pretend to solve, but the fact was very curious, as showing that bergs could touch the bottom without sweeping everything before them. I am sure that every geographer will be interested in the following fact, for it related to the late Sir John Ross, who stated that in his voyage in 1818 he brought up mud from 1050 fathoms, the deepest soundings ever obtained at that time; and in the mud were shells and fish. So to that old navigator they first owed that the fact of animal life in great depths had been discovered. We were talking the other day about the difficulties with regard to the intense cold and the aurora borealis affecting the telegraph. It might be interesting for the Society to know that a few seasons ago, Captain Kellett and Captain M'Clintock were beset in the ice in $74\frac{1}{2}°$ north, and they communicated with each other by telegraph from ship to ship. Now $74\frac{1}{2}°$ north latitude was a long way in the Arctic zone, and the cable, the practicability of which they were at present discussing, would lie a long way without it. That telegraph was at work throughout the whole winter, and was not in any way affected by that terrible bug-bear the aurora borealis, nor by the intense cold experienced so far north. (Applause.)

Mr. JOHN BALL: It is impossible to over-rate the importance of the subject brought to the attention of the Society by the gentlemen who read their papers. It is very desirable that we should not under-rate the difficulties which still remain, and which I think have not all of them been sufficiently brought forward, and it is by pointing out the difficulties at this early stage that we shall soonest arrive at the solution of those difficulties and the great end which I trust we shall live to see accomplished. I have long felt that this northern route was the right one; but I am not blind to the many difficulties which still remain in our way. With reference to the precise points which come particularly under the attention of this Society, and were brought clearly before us at the last meeting, there are two or three on which I wish to make a few observations. As to the Færöes, I apprehend there is no great difficulty. As to Iceland, it is proposed to land the cable at the south-east side of the island, to carry it across the island, and start again from Reikiavik in order to communicate from thence to Greenland. Now there, are we not unnecessarily encountering very serious difficulties and incurring a great increase of expenditure? It is true that it is possible to cross Iceland—perfectly possible—but so experienced a traveller as Dr. Rae found that it was not practicable without very great sacrifice of time, and encountering great difficulties, to take that direct route pretty nearly from east to west; but has he at all considered what the difficulties and cost of laying down a telegraphic wire are? I think he has mentioned—every traveller in Iceland has mentioned—that large district in the interior of the island over which you must hurry as fast as your horses can carry you, for the simple reason that there is nothing upon which to live; and if you are to lay down a telegraph across that interior desert of Iceland, you must be prepared for a vast number of difficulties and an enormous amount of money, and time, and labour, while they are doing it, and to accomplish it within a reasonable time. But, furthermore, how is it to be done? If you require piles, you must carry every pile into the island, across the island by horses, and supposing you could hire every one of the horses, there are not enough to enable you, in one or two seasons, to accomplish the work. And then, again, I have not alluded to the difficulty which every voyager will have encountered with respect to the breakers. Would it not be better to confine the difficulties to one point alone,—the entrance into the Faxe Fiord? I don't know whether it is proposed to land it at that point or not, but let the difficulties be confined to one point instead of two. With reference to Labrador, even though it be true that we can carry the 150 fathoms depth near to the coast, still that, my lord, may be as formidable as ten, or twenty, or fifty miles in another place. If it be a part of the coast on which there are icebergs, it is necessary, it appears to me, not only to carry your 150 fathoms of water close in shore, but you must get it safe behind some headland which will throw aside the icebergs. May it not be possible that by making the line a little longer, and running it into the strait of Belle Isle, the true solution will be found?

Sir Roderick Murchison:—If any Arctic geographer or voyager had been prepared to rise, I would not have occupied the time of the Society for one moment, upon a subject on which I know little. I will only say I am one of the very many well-wishers to every expedition of this sort which the Royal Geographical Society holds within its number. My lord, we are not here, as you have very properly said, to discuss the engineering merits of the project, nor are we really capable of estimating the difficulties that have been apprehended by my friend Mr. Ball; but we are capable of estimating, in a very decisive manner, the facts that Sir Leopold M'Clintock and Captain Sherard Osborn, who are men of great experience as Arctic voyagers, have pronounced in favour of this scheme.

I make no observations upon the scheme, except to say that had I risen at our last meeting, when your lordship adjourned the discussion to this evening, I should have felt it necessary to say that we owe a deep debt of obligation to those five gentlemen, who brought before us on that occasion their well elaborated and well received papers. They brought before us a mass of geographical knowledge which we never could have possessed, had not this expedition been matured—I say it to his honour—through the patronage of Lord Palmerston, and carried out through the active enterprise of the gentlemen who undertook this service.

Being in favour of all such expeditions, the moment my distinguished friend Captain Allen Young offered to go out in command of the little "Fox," in which you know I took a deep interest, I went down to Southampton, to attend a great meeting assembled by the Mayor in honour of this expedition. There Colonel Shaffner and Mr. Croskey, and the other gentlemen associated with them, and the Danish officers and others present received from us as Englishmen every encouragement we could offer to promote this enterprise.

My lord, the details of the two expeditions have been so admirably laid before you that I, for one, should have been contented with the results, if I were in any way embarked as a speculator in the success of the undertaking. But not having any pecuniary interest in it, or in any rival speculation, I may say to the gentlemen of the North Atlantic Company, you have put your case so well before the public, that no gentleman associated with other lines can say that it is possible to have treated the subject in a more ingenuous and fair manner. (Hear, hear, and applause.)

And now, my lord, I cannot sit down without saying, it gave me very great satisfaction to hear Dr. Rae expatiate, as he did, upon the warm and hearty reception that he received from all the inhabitants of the Danish settlement of the Færöe Islands and of Iceland. I say this in the presence of the Danish minister; and I am sure Dr. Rae was justified when he spoke in the name of Englishmen, of our gratitude to the King of Denmark, and to the government of that country, for having so warmly assisted us in one of those great enterprises, which may lead us to connect America with our own country. (Applause.)

Dr. Rae: My Lord,—The time allowed me at the last meeting of this Society was necessarily so short, that I left unmentioned some important points connected with my journey over Iceland; so I now beg to say a few words in reply to the observations of Mr. Ball.

The interior of Iceland over which we travelled is not devoid of grass, there being sufficient for our horses throughout the whole route, which offered comparatively few difficulties. Although two of the party were heavy men, we crossed the island in thirteen or fourteen days, riding the same horses at the rate of more than thirty miles per day. These ponies can carry a load of 200 lbs. or upwards, and pieces of wood ten or fifteen feet long can be carried by them over the highest and steepest ground on the route. The best line for the telegraph lies to the south of that we travelled over, and is marked on the chart by a dotted line. Along this route there is abundance of water, and grass also, except at one station, to supply which a depôt of hay could be made. This portion of the way (about the centre of Iceland), called the Springisandr road, is almost a flat table land. Again, some of the larger rivers that flow northward appear to be navigable for boats similar to those used in Hudson's Bay, which carry from three to four tons weight. By this means materials could be deposited at several stations near to the proposed line. I do not profess to be a surveyor, but having travelled a good deal both in the United States and Canada, I have seen telegraphs carried over more difficult and rougher country than we encountered in crossing Iceland. But should objections be offered to carrying the telegraph completely across the island, I am of opinion, from information obtained, that Portland, in the south of Iceland, although no safe harbour for ships, would form a good landing place for a cable. From Portland to Reikiavik the distance is not over ninety miles, and the road offers no obstacles. The object of our journey was not to make a minute or detailed survey, but to determine if it were practicable to carry a telegraph over the island. (Applause.)

APPENDIX.

Opinions of Greenlanders, on the practicability of laying the Telegraphic Cables upon the Coast of South Greenland.

The following extracts are taken from letters written in Greenland by residents there, who have had much experience upon the coast. They prove that the floe-ice is seen off the coast of Julianshaab in January or February, and that sometimes during the summer months there is none off the coast, and that every year, from and after the month of August until January, "the coast is free from ice." During the months that are free from ice a steamer can enter or depart with the cable, and lay it without the slightest interruption. The floe-ice runs north, along and within sight of the coast, unless driven to sea by extraordinary gales, which circumstance is always known. When it is driven to sea it clusters together in small patches of a few hundred yards in length.

The year 1860 was a remarkable era; there was a greater quantity of ice, and it remained longer upon the coast, than ever before within the memory of man or traditional accounts. Even during this year the cables could have been safely landed in Julianshaab and other fiords. Severe seasons occasionally occur in all climates, and so it has been in England the present winter (1861). The first fourteen days were colder than has been experienced since 1839, and it has been the third occurrence of severe cold within this century.

Extracts from a Letter dated Godthaab, Greenland, September 4th, 1860, from Dr. Rink, Inspector (Governor) of South Greenland.

"I need not tell you how interesting your letter from Julianshaab, of

October, last year, was to me, and how much I regretted the impossibility of meeting you.

"I have thought much on your grand enterprise, and though I doubt very much that the telegraph line can be carried across Greenland from the east side, I am convinced that it can be landed in some of the inlets (or fiords) in the vicinity of Cape Farewell, without fear of interruption from ice.

"The drift ice has been very bad this summer, and has deprived me of an opportunity to meet you."

Extracts from a Letter dated Julianshaab, November 2nd, 1860, from Mr. Möller, Governor of the Julianshaab District, a resident of Greenland since 1825.

"As a general rule the first drift ice appears in the months of January or February, but it is seldom of such extent as to reach further north than this district. When it does so, it is in the spring months.

"The Spitzbergen ice this year (1860) has beset the coast of this district from the month of January, but at some times not so much as to prevent vessels coming into this port.

"The worst months were June and July and a part of August. The reason that Col. Shaffner's vessel was able to run into this port so successfully in the autumn of last year (October 2nd, 1859), was that the ice had long before entirely left the coast; in fact the ice had gone away in August.

"The Colonel's ship, without the slightest hindrance, navigated the coast in the month of September, and put to sea from near this harbour in October 1859. But this year the ice has remained upon the coast longer than ever before within my memory. The coast is usually free from ice during the autumn months."

Extracts from a Letter dated Frederickshaab, October 11th, 1860, from Mr. Twede, Governor of the Frederickshaab District, a resident of Greenland since 1850.

"The ice off this coast has continued later this year than it has ever before during my residence in Greenland. During the past ten years, and before this season, the ice on the coast of this and the

Julianshaab Districts, with but one exception, has never prevented the Julianshaab vessels from sailing direct to this place, and they have generally arrived by the middle of August, and in 1855 on the 5th of July. Last year the vessels proceeded direct from sea to Julianshaab, the ice having disappeared off the coast early in the summer.

" In September 1855, I sailed along the coast of Julianshaab District beyond Cape Farewell to Itiblik, without having encountered any floe ice whatever. In September 1859, Col. Shaffner's ship was seen steering along the coast, and subsequently anchored at Kaksimiut, in the Julianshaab District, without any interruption from ice. The coast during the autumn months is nearly always free from ice."

Extract from a Letter dated Frederickshaab, October 11th, 1860, from the Rev. F. P. Barfoed, a resident of Greenland since 1852.

"During my residence in Greenland, I have never seen so much ice as during the present year. Dr. Rink, the Inspector (Governor) of South Greenland, and all other Danes and Greenlanders with whom I have consulted on the subject, are unanimous in declaring that never before within the memory of man has there been a year of so much ice as that of 1860.

"It is my opinion that both the east and the west coasts are every year, at times, free from ice, but possibly not on both coasts at the same time; and it is my decided opinion that a steamer will almost always find an opportunity to proceed unobstructed, direct from and to Julianshaab District, either from or to Labrador and Iceland, after, and many times during, the month of August. In the month of September 1859, I saw Col. Shaffner's vessel coasting southward, and in a few days afterwards we heard that he had anchored at Kaksimiut in the Julianshaab District. There was but little ice last year, and in an especial manner was the Julianshaab District free from ice."

Statement of Mr. Motzfeldt, Superintendent at Kaksimiut, one of the oldset residents in Greenland, November 1860.

"The year 1860 has been one of the worst ice years that we have ever had. Last year (October 2nd, 1859), Colonel Shaffner's ship entered Kaksimiut without difficulty. In the months of September,

October, and November, generally, this coast is free from ice, and a steam vessel can enter from the sea and depart from the coast almost any day, and not be interrupted during the autumn months. I am now over fifty years of age, and do not remember ever having seen an ice year as bad as this (1860). Even this year the coast of the Julianshaab District has been free from ice during the past and present months."

Extract from a Letter dated Julianshaab, November 3rd, 1860, from J. Anderson, Captain in the Royal Greenland Commercial Service since 1848.

"During eight years, I was boatswain to a merchant vessel trading to this district, and during the last twelve years commander of the district schooner 'Activ,' of 22 tons burthen. During those periods, I have as an old sailor (for I have been forty years at sea) made many careful observations as to the nature of the ice along this coast. I have also kept a correct log (book which is now in the possession of the Board of Management of the Royal Danish Commercial Institution of Greenland), ever since the year 1849.

"Having gone through that log book, and compared it with my own views, I now express the opinion that, as a general rule, the ice mostly lies on the coast during the months of April and May, and that, on the other hand, the months which are free from ice are September, October, and November. It may be taken as a general rule, that during the months of September and October, even in what are called "bad ice years," this coast is free from ice.

"My own experience enables me to state positively—and my log book will bear me out in the statement—that steam-vessels can run straight into the settlement from the sea, with certainty, during the months before mentioned; and, as a proof of the correctness of this opinion, reference may be made to this very year, which, in regard to ice, to my own knowledge, has been the worst and the longest in duration of any 'bad ice year' since my arrival here in the year 1841. The large vessel, 'Bulldog,' came straight to this place from sea; and since the date of its arrival, up to the present time, any ship, be it what it may, might have been equally able to make this port, from the sea, and to leave it again without interruption from ice.

"Last year (1859) an American barque came straight into Kaksimiut, which is a little to the north of this settlement; but that is not to be wondered at, as last year was what we call here an 'ice-free year;' and if the captain of that vessel had been acquainted with the coast navigation, he might have easily come in and gone out more than twenty times, and from as many places."

Extract from a Letter dated Julianshaab, October 23rd, 1860, from Mr. Dorph, retired assistant from Nennortalik; I. Holm, superintendent at Sardlok; M. Schonheyder, superintendent at East Proven; and S. G. Lund, superintendent at South Proven.

"In compliance with the request made to us, we hereby declare that the present year has been what is called, in Greenland, a 'bad ice year.'

"The large ice remained an unusually long time at the shore. Neither of us, nor any of the old Greenlanders with whom we have conversed on the subject, remember ever having seen, in the course of our lives, the ice remaining for so long a period along the coasts.

"The ice lay so thick for so long a time that the hunters were prevented thereby from performing their usual work, even in their Kayaks. In the present year the ice set in from the first month, and remained until September, when it disappeared. This coast is usually free from ice during the autumn months, and vessels can enter and depart without interruption from ice.

"One of the undersigned, Mr. Dorph, who has resided in this district during forty-one years, remembers well the year 1825, which was a remarkable 'bad ice year,' but it was not such a 'bad' one as the year 1860."

Extract from a Letter dated Frederickshaab, October 12th, 1860, from T. Ordrorpe and C. Fotel, Captains of vessels from Denmark.

"On the 19th of June (1860), at about eight miles from land off Cape Farewell, we found much ice; on the 22nd, off Julianshaab, there was no ice. The ice was in patches, more north, and the greater part of our voyage on the coast, it was near to land. We were informed by the old people of Godthaab, that during the past thirty years there has not been any year with as much ice on the Greenland coast as the present, 1860."

Opinions of Lieut. von Zeilau, of the Royal Danish Army, and Arnljot Olafsson, Member of the Icelandic Diet (both of whom accompanied the "Fox" Telegraph Expedition in 1860), Commissioners appointed by the Royal Danish Government, on the practicability of the proposed telegraphic route.

"We concur in the opinion that it is practicable to submerge or land telegraph-cables at the places selected on the Færöe Islands, and also to construct a line over-land there, and that in our opinion, cables, when landed, will remain undisturbed.

"We have the same opinion in regard to the landing and maintaining cables on the coast of Iceland, and also that a line can be constructed across Iceland with all reasonable prospects of permanency.

"As regards Greenland, we give it as our decided opinion, based upon our own observations and received information from the best authorities, that cables can be safely laid from Greenland during the autumn months, for instance, at the Igalikko (Julianshaab) Fiord, where cables can be submerged beyond the reach of icebergs."

THE ROYAL DANISH CONCESSION
FOR AN
ELECTRIC TELEGRAPH
BETWEEN
EUROPE AND AMERICA, VIÂ FARÖE ISLES, ICELAND, AND GREENLAND; AND THE OFFICIAL CORRESPONDENCE.

WE Frederick VII., by the Grace of God, King of Denmark, of the Vandals and Goths, Duke of Schleswig, Holstein, Stormarn, Ditmarch, Lauenburg and Oldenburg,

Do Hereby Make Known,

That in accordance with the most humble request and supplication which has been made to us by Tal. P. Shaffner of the State of Kentucky, of the United States of North America, concerning the project herein mentioned, we have been most graciously pleased, by these presents, to permit the said Shaffner to construct a Telegraph from North America to Copenhagen, upon the following stipulations.

I.

To Tal. P. Shaffner of the State of Kentucky, United States of America, is herewith given permission to construct a Telegraph line, at his own cost and risk, upon the territory of the kingdom of Denmark, from North America to Copenhagen, viâ Greenland, Iceland, the Faröe Isles, Norway and Sweden, and from thence to Copenhagen, the only ending point.

The said Telegraph line shall connect with the Royal Danish Government Telegraph lines at or near Copenhagen, or Elsinore, upon the Island of Zealand, or to connect at any other point upon the said Island which may be agreed to by the Government of Denmark, at the expense of the said Tal. P. Shaffner.

The Danish Government reserves to itself the determination as to the conditions of landing the projected Telegraph line upon the Island of Zealand and the section of line built thereon; and also the administration as to business connection between the herein-projected Telegraph, and the Royal Telegraph lines.

II.

The said Tal. P. Shaffner may expect that the lands required for the purposes of the herein-named project in the territories of Greenland, Iceland, and the Faröe Isles, shall be made over to him, whether belonging to the Crown, or to private persons, upon such conditions as shall hereafter be determined upon.

III.

The said Tal. P. Shaffner binds himself to finish the before-mentioned undertaking as soon as possible; and it is accorded to him the term of ten years, from the date of this Concession, for the completion of the said Telegraph. If the said project be not finished and in operation within the said time, then the Danish Government shall be entitled to cancel this Concession without any obligation to make any reimbursement whatever to the said Tal. P. Shaffner.

IV.

When the said Telegraph line, or a part thereof, shall be completed, then it shall be permitted to the said Tal. P. Shaffner to use the same for the transmission of dispatches, insomuch as the contents thereof may not be regarded as dangerous to the Danish State, or the common weal, from and to all nations; and the Danish Government will bestow all necessary care, vigilance, and means which may be within its command to ensure the free, impartial, and unhindered use of the said Telegraph line.

V.

It is conditioned that whenever the said Telegraph line shall traverse Royal Danish land or sea territory, the owners of the said Telegraph and their agents shall be subjected to the laws of the Danish Government, and to such judgments and resolutions as shall be enacted by the Judicial Tribunals of the Government, in every way as if they were Royal Danish subjects, forbidding hereby any ex-territorial rights or other exceptions to the decrees as aforesaid.

VI.

On the completion of the herein-named project to Copenhagen, then the Danish Government will assume the responsibility to secure the necessary celerity in the transmission of all dispatches, from and to North America, over the Royal Danish Telegraph lines.

VII.

When all the States, through the territories of which the herein-projected Telegraph traverses between North America and Europe, shall reduce the tariff of charges, for the transmission of dispatches going to or coming from the herein-projected Telegraph line, then the Danish Government will, on its part, allow a corresponding reduction.

VIII.

The Danish Government reserves to itself the free transmission of Government dispatches upon the Telegraph line herein projected, between the ends thereof, or to any station on the said line, to the amount of three hundred words per month, and such dispatches shall have preference over private dispatches.

IX.

As a guaranty to the said Tal. P. Shaffner from losses in the construction of the herein-projected Telegraph, and the expenses connected therewith, it is hereby given him the exclusive permission for the use of the said Telegraphic line from North America to Copenhagen, agreeable to Article IV., for the term of one hundred years from the date of this present Concession, with the exclusion of any and all other undertakings of the same kind, or nature, on Royal Danish territories, save that which the said Tal. P. Shaffner intends to establish, either by following the direction laid down in this present Concession, or by the establishment of branch lines to form another telegraphic connection between Europe and North America.

The Danish Government reserves to itself the right, after the lapse of the said specified time, to grant a new Concession to whom it wishes, and on such conditions as it may then determine upon; though those to whom this Concession may then have been transferred shall have the preference; provided they will agree to the conditions then to be made in the premises by the Danish Government.

X.

For security that this project be perfected with the necessary energy and dispatch, Tal. P. Shaffner binds himself, after the expiration of three years from the date of this Concession, to produce evidence to the Danish Government that he has expended towards the herein-named project at least the sum of 100,000 dollars, Danish currency. If such an amount shall not have been expended, he binds himself then to deposit with the Danish Government any such difference, in cash, and such an amount shall be paid back to him only when the said enterprise has been finished and brought into actual operation, within the time of ten years as named in Article III. In the contrary case, said amount shall belong to the Danish Government. Should he, in the above-named case refuse or avoid to deposit the difference of amount, then shall this Concession be deemed, without any further ceremony, as retracted, without the right of Tal. P. Shaffner to demand any repayment.

XI.

After the completion of the Telegraph line herein projected, if there shall occur an interruption of communication over the same, the said Tal. P. Shaffner binds himself to re-establish the communication in the best manner, and as soon as possible; but, should such interruption continue for a longer time than ten years, then the Danish Government shall have the right to cancel the Concession without any reimbursement, as mentioned in Article III.

XII.

Whenever it shall be deemed necessary or desirable to carry the said Telegraph line in another direction than the one mentioned in Article I., or that the same shall branch out in such a way that Copenhagen will not longer remain the only ending point of the said Telegraph line, the said Tal. P. Shaffner may then expect the permission therefor, if he shall agree with the Danish Government about the way, means, and time for the completion and execution of such a proposition.

XIII.

The said Tal. P. Shaffner shall have the right to transfer the herewit given Concession to whom he pleases, to a company or to an individual, with the same privileges and obligations which are binding upon him; but all such transfer shall be immediately brought to the knowledge of the Danish Government, and he or they who may own the control of the undertaking, during the time that the work may be done and after the completion of the same, shall be obliged to give to the Danish Government, after a suitable lapse of time, all information which the said Government may demand on all matters regarding the said enterprise.

And the said Tal. P. Shaffner shall in all cases, wherein the interpretation of words, or the exact meaning of this Concession, whereof a question may arise, be subjected to the decision of the Minister of the Interior.

WE COMMAND herewith each and every person or individual not to create obstacles to the above decree.

GIVEN at Scodsburg, on this the 16th day of August, 1854, under our Royal hand and seal.

FREDERICK R.

L. S.
R.   TILLISCH.

CONCESSION to Tal. P. Shaffner, of the State of Kentucky, United States of North America, for the Establishment and use of a Telegraph from North America to Copenhagen.

I, the undersigned, Nicholas Christian Levin Abrahams, Knight of the Order of Dannebrog, and decorated with the Silver Cross of the same Order, Knight of the Legion of Honour, and of the Polar Star, sole Notary Public Royal in and for this city and royal residence of Copenhagen, do hereby certify that the preceding is a true translation of the original Royal Concession, written in the Danish language.

IN TESTIMONY whereof, under my Notarial Firm and Seal of Office.

Copenhagen, the 27th of February, 1860.

N. C. L. ABRAHAMS,
Not. Publ.   L. S.

[COPY.]

To HIS EXCELLENCY THE MINISTER OF FINANCE
OF THE KINGDOM OF DENMARK.

ON August 16th, 1854, I obtained from His Majesty the King of Denmark a Concession, with the exclusive right of building a Telegraph over Greenland, Iceland, and the Faröe Islands, to establish a telegraphic communication between Europe and North America.

The 10th Article of this Concession required that after three years I should produce evidence that I had expended not less than 100,000 rix dollars, Danish currency, towards the enterprise. Should the amount expended towards the project not reach 100,000 dollars, the balance was to be deposited with the Danish Government.

I have in reality, before the lapse of the three years, spent more than double the amount; but I regret to hear that the Government has not considered the evidence produced by me as satisfactory, as regards form, and that the Government, while convinced that I have sacrificed a great sum of money for preliminary purposes towards the project, even on account of the informality of my evidence, has found it impossible to judge what amount in reality has been expended.

With a view of avoiding the delay of the great work consequent upon the uncertainty whether or not the Government will consider Article X. fulfilled by me, I propose to deposit the whole amount of 100,000 dollars, in order to put an end to the question. It should be understood that it is a further sacrifice on my side in this way to add to the amount already expended, and the necessary outlays for furthering the project; but on the other hand, I am so fully convinced of the whole line being finished even *before* the stipulated term, that I consider the deposit of but short duration.

As the money to be deposited must be gotten in North America, I request a delay of three months from to-day, hoping that the Government will accept this with a view of reaching thereby a final settlement so much desired on both hands.

My request is, therefore, that an arrangement be made with regard to the fulfilment of Article X. in the Concession given me on the 16th of August, 1854, so that the said Article shall be considered fulfilled by me when, on or before the 20th of March, 1860, I shall have deposited with the Danish Government in Copenhagen the sum of 100,000 dollars, Danish currency, whereas the Concession, if this be not done on or before the said day, shall be without question considered as forfeited.

Very respectfully,
Your Excellency's most humble
and obedient servant,
(Signed) TAL. P. SHAFFNER.

COPENHAGEN, *Dec. 20th*, 1859.

[COPY.]

Translation from the Danish Language.

MINISTRY OF FINANCES.

COPENHAGEN, the 21st *Dec.*, 1859.

UNDER date of yesterday, I have received your letter of the same date, wherein you solicit a delay until the 20th March next year, for the payment of an amount of 100,000 R. Dlrs., Danish money, which sum, according to Article X. of the Concession granted to you under date of 16th August, 1854, for the establishing of a telegraphic line between North America and Denmark through Greenland, Iceland, and the Islands of Farõe, you have engaged yourself to deposit with the Royal Danish Government, acknowledging that in case the said obligation be not fulfilled at the said epoch, the beforesaid Concession, according to its contents, is to be forfeited.

In answer to your said letter, it is hereby communicated to you that, with respect to circumstances, the Minister of Finances approves of the said delay for the payment of the above-mentioned sum of 100,000 R. Dlrs. Danish money, being granted to you: however, this said amount not being deposited with the Royal Danish Government, by payment made to the Ministry of Finances in Copenhagen, on or before the 20th March next year, the Concession granted to you the 16th August, 1854, will be considered as forfeited by you, without the Danish Government being bound to give you any notice thereof. If, on the contrary, you fulfil the before-mentioned engagement, the stipulations of Article X. of the Concession will be considered as fulfilled, and in consequence thereof, provided the remaining conditions of the Concession be exactly observed, the said Concession will have its full force.

    (Signed)   REGNAR WESTENHOLZ.

To Mr. TAL. P. SHAFFNER,
 Of Kentucky, United States of America.

                  RAMUS.

[COPY.]

  To His EXCELLENCY
   MR. WESTENHOLZ,
    MINISTER OF FINANCE, &c., &c.

In conformity with your Excellency's letter of the 21st December, 1859, relative to the deposit of 100,000 dollars, Danish currency, as therein mentioned, I have arranged with the private Bank in Copenhagen to deposit the said sum of money to the credit of the Government, and I desire to deliver to your Excellency the bank certificate of deposit in payment of the said 100,000 dollars, Danish, if the same be acceptable to your Excellency.

I request that the payment shall be upon the condition, that if it shall be found by me, or my associates, that the projected telegraph cannot be constructed, owing to physical difficulties with reference to Iceland and Greenland, then the money deposited as above stated shall be returned to me or my associates. And, that we shall have one year, from this date, to make the necessary explorations and surveys to determine the practicability of constructing the projected telegraph.

I request further, that the Danish Government will appoint an assistant to accompany the expedition, at our expense, to aid in determining the practicability of the proposed telegraphic route.

The time and place of embarcation will be made known to the Government at the earliest moment possible, so as to permit a selection of a proper person for the specialty.

I would respectfully request your Excellency to accept of the foregoing considerations.

       I am, very truly,
        Your Excellency's most humble
         obedient servant,
            TAL. P. SHAFFNER.

COPENHAGEN, *Feb. 13th,* 1860.

[COPY.]

     Translation from the Danish Language.

      MINISTRY OF FINANCES.

        COPENHAGEN, the 16th *February,* 1860.

The Ministry acknowledges receipt of your honoured letter of the 13th inst., and in answer thereto communicates as follows.

The Ministry has no objection to the fulfilling of your obligation—in order to maintain the Concession granted to you concerning a Telegraphic

Establishment between Europe and North America—by depositing the sum of 100,000 r. d., Danish money, in such a manner that the aforesaid amount is by you deposited in the private Bank of Copenhagen, so that this amount may be to the disposal of the Danish Government, and consequently the Deposit Bond to be delivered by you to the Ministry of Finances.

The Ministry has no objection to promising you, that provided the surveys you are to undertake in order to find out the possibility of a telegraphic connection from North America to Europe, over Greenland, Iceland, and the Isles Faröe, prove that the enterprise, on account of physical circumstances, is impracticable, and this be proved to the Danish Government by satisfactory evidences, amongst others by testimonials given by the men designed, as hereinafter stipulated by the Danish Government, to partake of the researches, the beforesaid amount will be repaid to you, and the Concession granted to you will consequently be nullified.

Neither will the Ministry have any objection to allowing to you and your eventual partners one year from this date for examining and surveying the practicability of the projected enterprise.

The Ministry will fulfil your wish that Delegates from here may, on your and your partners' expenses, partake of the intended expedition; only the Ministry must think proper that in the interest of the undertaking *two* persons be sent from here to partake in the said expedition; and the Ministry reserves to itself, further, to nominate the persons proper to be delegated for this purpose, and, according to your offer, you are to notify in time to the Government the epoch at which the expedition is to take its commencement. The Ministry reserves, also, to itself to stipulate the pecuniary conditions, &c., under which are to be placed the said Delegates, if an agreement should not be made between you and the persons in question.

As for the rest, provided you will give timely information on that subject, the Ministry will take all convenient measures that the Royal authorities on the Isles of Faröe, Iceland, and Greenland, afford you all possible help and assistance.

REGNAR WESTENHOLZ.

To TAL. P. SHAFFNER, Esq.

RAMUS.

[COPY.]

To HIS EXCELLENCY
    MR. WESTENHOLZ,
        MINISTER OF FINANCE, &c., &c.

For the purpose of being the better able to accomplish success in the completion of the telegraph from North America to Europe, as contemplated in the Concession awarded to me by his Majesty the King of Denmark, on the 16th day of August, 1854, I herewith petition for the following amendments or alterations to the said Concession, viz.—

I.

Instead of constructing the projected telegraph line direct from the Faröe Isles through Norway and Sweden, I shall have the right to construct two branches of the said telegraph from the Faröe Isles to Europe.

    1st. One branch of the projected telegraph from North America to be extended from the Faröe Isles to Scotland, either direct or *viâ* the Shetland or Orkney Isles.

2nd. The other branch of the projected telegraph to be extended to the coast of Jutland in Denmark, the point of landing to be determined by an agreement hereafter to be made by the Danish Government. This branch line is to run from the Faröe Isles to Norway, and thence to Denmark, provided the Danish Government can secure satisfactory arrangements from the Norwegian Government for the extension of the said branch line to the coast of Jutland in Denmark. If, however, the Danish Government cannot secure the satisfactory arrangements from the Norwegian Government, or if, for other reasons, it shall prefer to run the line direct from Scotland to Denmark, I am, in that case, to construct the line from Scotland direct to Jutland in Denmark.

The transmission of dispatches over the respective branch lines to be as following, viz.—Over the Scotland branch are to be sent dispatches to and from Great Britain and Ireland; and over the Norwegian branch line are to be sent the dispatches to and from the Continent of Europe.

II.

That the completion of the Scotland branch line of the projected telegraph on or before the expiration of the ten years mentioned in Art. III. of the Concession of August 16th, 1854, shall be considered as a fulfilment of the Concession; provided, however, that within two years after the expiration of the said ten years, the branch line to Denmark shall be completed.

III.

In case the branch line from Scotland direct to Jutland be authorised, I am to assume all responsibilities arising from any infringement upon the rights of other concessioners.

IV.

In case of dissatisfaction arising from the decision of the Minister of the Interior as to the interpretation of any of the clauses of the Concession and amendments thereof, I shall have the right of appeal to the Council of Ministers.

With the foregoing amendments to the Royal Concession granted to me on the 16th August, 1854, I shall confide in the hope of an early consummation of the great enterprise. If the granting of the amendments above given shall require much delay as to formality, I will thank your Excellency for such a recognition of their importance immediately that shall serve the purpose of securing the confidence of the public, that they will be formally accorded at such time hereafter as the interest of the enterprise may demand.

I would respectfully request that your Excellency will give this petition the earliest attention possible, so that I may be able to enter upon an active service of the telegraph immediately.

I am, very respectfully,
Your Excellency's most humble
and obedient servant,
TAL. P. SHAFFNER.

COPENHAGEN, *Feb. 16th*, 1860.

THE ROYAL DANISH CONCESSION.

[COPY.]

Translation from the Danish Language.

MINISTRY OF FINANCES.

COPENHAGEN, the 3rd of *March*, 1860.

The Ministry has received your letter of the 16th last month, and in answer thereto communicates as follows :—

When you have fulfilled the obligation incumbent on you with regard to the depositing of a guarantee sum with respect to the Concession granted to you for a Telegraphic Establishment between Europe and North America, and when the practicability of such an establishment has been proved according to what has been expressed in the letter of the Ministry to you of the 16th last month, the Ministry of Finances will be willing to represent to His Royal Majesty, that in the Royal Concession granted to you the 16th of August, 1854, the following changes be made :—

>That instead of the telegraph line, according to the said Concession, being conducted from the Faröe Isles directly over Norway and Sweden to Copenhagen, it will be permitted to you to conduct a line from the Faröe Isles to Scotland ; also, that you, if you deem it necessary, may land the line on the Shetland Isles and the Orkneys, on condition that, in the same time, a line be conducted to Jutland, to a point more precisely to be determined upon on agreement with the Danish Government. It is reserved to the Danish Government the determination as to the conducting of the said line either from the Faröe Isles over Norway to Denmark, or from a landing-point in Scotland to Denmark. If the first-named line be chosen, the Danish Government undertakes to procure the necessary permission from the Royal Norwegian Government.
>
>That the line from the Faröe Isles to Scotland be only used for correspondence between North America on the one side, and Great Britain and Ireland on the other side, consequently not for the correspondence between North America and the Continent of Europe, that correspondence being exclusively reserved to the line conducted to Denmark, whichever of the two said ways this line may take.
>
>That all questions relative to the interpretation of the words and true meaning of the Concession granted to you, if so wished by you or the other concessioners, shall, through the Ministry concerned, be laid before the King's Privy Council.

Finally, it is observed, that the Ministry accepts your offer to undertake the removing of such hindrances as might present themselves against the beforesaid line between Scotland and Denmark arising from the pretensions of other concessioners.

(Signed)   FENGER,

To TAL. P. SHAFFNER, Esq.

RAMUS.

[COPY.]

Translation from the Danish.

OFFICE OF THE MINISTER OF FINANCE.
COPENHAGEN, *March* 20*th*, 1860.

Whilst having the pleasure of hereby acknowledging the receipt of the Bill of Exchange for 100,000 rix dollars at sight on the National Bank, and forwarded in your esteemed favour of the 15th of this month, I must not omit adding, that the caution money or security which by the terms of the Concession accorded on the 16th of August, 1854, Colonel Tal. P. Shaffner was bound to deposit, with reference to the laying down of an Electric Telegraph between Europe and North America, is now fully paid in.

(Signed)   FENGER,

To Messrs. CROSKEY & Co.,
    84, King William Street, City, London.

MASTANER.

www.ingramcontent.com/pod-product-compliance
Lightning Source LLC
Chambersburg PA
CBHW030410170426
43202CB00010B/1552